Digital Media

Postphenomenology and the Philosophy of Technology

Editor-in-Chief
Robert Rosenberger, Georgia Institute of Technology

Executive Editors
Don Ihde, Stony Brook University, Emeritus
Peter-Paul Verbeek, University of Twente

Technological advances affect everything from our understandings of ethics, politics, and communication, to gender, science, and selfhood. Philosophical reflection on technology helps draw out and analyze the nature of these changes, and helps us understand both the broad patterns and the concrete details of technological effects. This book series provides a publication outlet for the field of the philosophy of technology in general, and the school of thought called "postphenomenology" in particular. Philosophy of technology applies insights from the history of philosophy to current issues in technology, and reflects on how technological developments change our understanding of philosophical issues. In response, postphenomenology analyzes human relationships with technologies, while integrating philosophical commitments of the American pragmatist tradition of thought.

Titles in the Series

Digital Media: Human–Technology Connection, by Stacey O. Irwin

Acoustic Technics, by Don Ihde

A Postphenomenological Inquiry of Cell Phones: Genealogies, Meanings, and Becoming, by Galit P. Wellner

Postphenomenology and Technoscience: The Manhattan Papers, Edited by Jan Kyrre Berg O. Friis and Robert P. Crease

Postphenomenological Investigations: Essays on Human–Technology Relations, Edited by Peter-Paul Verbeek and Robert Rosenberger

Design, Meditation, and the Posthuman, Edited by Dennis M. Weiss, Amy D. Propen, and Colbey Emmerson Reid

Digital Media

Human–Technology Connection

Stacey O'Neal Irwin

Foreword by Don Ihde

LEXINGTON BOOKS
Lanham • Boulder • New York • London

Published by Lexington Books
An imprint of The Rowman & Littlefield Publishing Group, Inc.
4501 Forbes Boulevard, Suite 200, Lanham, Maryland 20706
www.rowman.com

Unit A, Whitacre Mews, 26-34 Stannary Street, London SE11 4AB

Copyright © 2016 by Lexington Books

All rights reserved. No part of this book may be reproduced in any form or by any electronic or mechanical means, including information storage and retrieval systems, without written permission from the publisher, except by a reviewer who may quote passages in a review.

British Library Cataloguing in Publication Information Available

Library of Congress Cataloging-in-Publication Data

Names: O'Neal Irwin, Stacey, author.
Title: Digital media : human-technology connections / by Stacey O'Neal Irwin ; foreword by Don Ihde.
Description: Lanham : Lexington Books, 2016. | Series: Postphenomenology and the philosophy of technology | Includes bibliographical references and index.
Identifiers: LCCN 2016005178 (print) | LCCN 2016005885 (ebook) | ISBN 9780739186534 (cloth : alk. paper) | ISBN 9780739186541 (Electronic)
Subjects: LCSH: Technology--Social aspects--Case studies. | Digital media--Social aspects--Case studies.
Classification: LCC T14.5 .O54 2016 (print) | LCC T14.5 (ebook) | DDC 303.48/33--dc23
LC record available at http://lccn.loc.gov/2016005178

∞™ The paper used in this publication meets the minimum requirements of American National Standard for Information Sciences Permanence of Paper for Printed Library Materials, ANSI/NISO Z39.48-1992.

Printed in the United States of America

For Faith, Daniel, and David, who feel the technological texture.

Contents

Foreword ix

Preface xi

I: Raw Materials 1

  1 Exploring the Texture: An Introduction 3

  2 Describing Digital Media 17

  3 Digging Deeper through Phenomenology 29

II: Feeling the Weave 47

  4 Case: The Screen 49

  5 Case: Dwelling in Digital Sound 63

  6 Case: Earbud Embodiment 77

  7 Case: Portable Sound 89

  8 Case: Dubstep 101

  9 Case: The Photo Manipulation Aesthetic 115

  10 Case: Data Mining 127

  11 Case: Aggregate News 141

  12 Case: Self-Tracking 155

Epilogue: Convergence: Revising the Texture 169

References 173

Index 181

Foreword

My favorite image of Stacey Irwin shows her seated in and operating a large excavator. She and her husband are constructing a new house and Stacey, as with everything she deals with, "gets into it." Postphenomenological research studies are now in their eighth year of being presented, published, and now in monograph and book form. The producers of this research are all, in different ways, *participant observers,* and the interdisciplinary groupings are now clearly emergent. There are the phenomenological philosophers, who upon developing sensitivities to materiality, usually begin to think of themselves as *postphenomenological.* Then there are the cultural anthropologists or ethnologists who focus upon the practices of science-technology, or better, technoscience practices. And there are the media-communication-informatics folk who may be the most *participant* of the participant-observers who make up this community. That is clearly Stacey. Add, finally, a growing group of performance artists and you have the mix. *Digital Media; Human-Technology Connection* romps through the very most popular and current of digital technologies in the media world. What is unusual about this volume is the intensity Stacey brings to her topics—she is "into them." She is a user, a practitioner of screens, earbuds, iPods, DJ and remix technologies, Photoshop, data mining, and aggregate news. She has been a user of all these technologies and been a DJ, listener, netizen, photo and journalistic editor. She unites the various digital technologies under the metaphor of *technological texture,* a neologism of mine, and adds to it weaving metaphors. These are most appropriate. I, myself, am a collector of handwoven rugs—from travels, many cultures, all over the world. And when these lie next to one another one can see that the woof and warp of weaving partially determines, though not strictly so, what the designs may be. Finer knotting and silks allow for differences from rougher wool or flax. These subtle differences can be seen whether the design is geometric, animal or flower motifed. There is a texture which runs through all of the variations. Stacey also does more with acoustic digital technologies than most writers. She traces some of the history of analogue to digital, notes how earbuds are embodied, and analyses the way in which iPods and other portable acoustic technologies insulate one from the ordinary worlds. In this way Stacey shows insight into today's youth lifeworlds which average some thirty-seven hours a week "plugged in." While most of the digital technologies engage individuals, even solipcistically, the contemporary contexts of big data and its outcomes—as in aggregate news—also create a multistable set of informations for our new worlds. Stacey understands the deep role of multistability in today's technologies. These, whether so designed as the multi-purposed mobile phone, or as unintended outcomes in data development, make for both unique patterns of contemporaneity and for our complex ambiguity which we cannot avoid. In this respect for complexity, Stacey clearly fits postphenomenology in its forms. Stacey also worries about outcomes and normativity. When I read her concerns over Photoshop and the presumed "perfections" possible for digital editing, as a painter I thought of my favorite "phenomenological" painters, Matisse and Picasso. Today's analytic technologies can show how many times they each changed and modified their own paintings. In the case of Matisse's "The Bathers" these changes went on for nearly thirty years before it was taken as finished. In Picasso's cases the deformations of his models are as radical as can be imagined.

Don Ihde
Distinguished Professor of Philosophy, Emeritus
Stony Brook University

Preface

Digital media is ever-present, its content is always available, and its context is ubiquitous. Walk down the street and you'll find people sitting alone or together, looking at the technological "face" reflecting back at them. In the café, on the bus, at an intersection, around the table, and in schools, workplaces, gyms and restaurants, people consistently use their devices. We are so technologically textured that it seems odd to think about what this human-technology experience is like. So many of us are surrounding ourselves with digital media technology to the point where we can not even think about life without it, or even think without it.

But the very ubiquity of the media is precisely why we need to critically think about and study this human-technology connection. What is our relationship with our technology? Yes, we might be experiencing it, but are we really paying attention to the experience? The ambition of this study is to consider the idea that digital media technologies and the content it creates and spreads does not lack effect. Digital media textures our world in multiple and varied ways. The intent is to investigate the connection we have to digital media content and devices through postphenomenology, and other associated philosophical underpinnings, to create a richer understanding of this lived experience.

The cover image for this book is a photograph of a NCR 551 printer memory buffer for the Series 500 Computer System, used from 1965 to the late 1970's. This type of printer was once installed in trailers and used by the United States Air Force in the Vietnam War. This particular photo was from a buffer now housed in my father's basement, where he has a computer museum of sorts, from his time as a computer business owner and computer teacher. I grew up surrounded by computer components and always thought that a buffer board looked like a weaver's loom. Computers and textiles share an intertwining because punch cards used in the mechanical Jacquard head dobby loom created a gateway for better understanding the punch card system used for early computer programming and data entry. The printer memory buffer is the perfect image to describe technological texture.

I am grateful for the rich tapestry of colleagues, friends, and family who have supported this work. My colleagues at Millersville University of Pennsylvania, both in the Department of Communication and Theatre and across campus, have provided me with much support and encouragement along the way. The opportunity to present my work at conferences has been foundational to this project and the Millersville University faculty grants committee supported portions of almost every conference presentation that contributed to this text and provided released time to complete the writing. Grants to travel to academic conferences are invaluable and make scholarship possible for those of us who are teacher-scholars. Broadcast Education Association (BEA), Media Ecology Association (MEA), National Communication Association (NCA), Society of Existential and Phenomenological Theory and Culture (EPTC), Society for Social Studies of Science (4S), Society for Phenomenology and Human Sciences (SPHS), and Society for Phenomenology and Media (SPM) conferences and conventions have allowed me to share my work and learn from colleagues.

I would especially like to thank my university colleagues; Greg Seigworth for his encouragement and enthusiasm and love for similar kinds of philosophy, Bill Dorman for consistently reminding me why we do what we do as college teachers, Rob Spicer for a close reading and significant feedback, Tom Boyle for departmental support, and Marlene Arnold and Tracey Weis for their encouragement in big and small ways.

This kind of project starts in small discussions at conferences and continues as debates over coffee (or wine) at the end of a long day of listening to research. I am indebted to my conference colleagues who have helped me formulate ideas around the cases discussed in this book. I feel very fortunate to find a mentor in Don Ihde, whose many books, presentations and strong use of metaphor have knit a texture into my scholarly writing. His encouragement and warm invitation into the Postphenomenological Research Group has allowed me to find a home for my research and for that I am forever appreciative. A big thank you to the Postphenomenology and Philosophy of Technology Series editors Robert Rosenberger and Peter-Paul Verbeek for working to bring this series to fruition and Yoni Van Den Eede for encouraging me to write. Additionally, meaningful philosophical conversation with Corey Anton, Isaac Catt, Kathryn Egan, Greg Faller, Anette Forss, Norm Friesen, Erik Garrett, Dave Koukal, Pieter Lemmens, Nicola Liberati, Lars Lundsten, Paul Majkut, Chris Nagel, Shogi Nagataki, Ellen Rose, Cynthia Rossi and Galit Wellner allowed me to better understand the direction of my work through the years. Special thanks to academic mentors Francine Hultgren, Barry Moore and Norma Pecora for teaching me about scholarship and academia.

A debt of gratitude and much love to my personal support system; My husband Dukie, my three children Faith, Daniel, and David, Judith and Bradford, Duke and Linda, the Graves, Jurgensen, and Hooley families, James Volpe for photo manipulation work, Daniel J. Irwin for the cover photo, and Mindy Kinsey von Schmidt-Pauli for editing and proofreading expertise. And finally, many thanks to the staff at Lexington Books, especially Jana Hodges-Kluck and her staff.

I

Raw Materials

Chapter One

Exploring the Texture: An Introduction

To be alive in today's world is to be technologically textured. Our phones, TVs, games and books come on little devices where novelists and game designers explore dystopian futures, where technology rules and online bloggers and early morning media programs extol the next new "thing" in our daily technologically mediated world. We can easily search for most any topic, situation or issue using a plethora of technologies or the one that is designed to do it all. Because, as we know, there is an app for that, or a GPS coordinate at the ready to get us "there," or a new technology altogether that is specifically designed for our needs and wants and interests. Immersive 3D environments and cognitive technologies bring connection from vague to realistic. It is time to power up and *pause* our digital media in light of its highly popular consumption by the world's first fully digital generation, to take up an investigation of exactly what might be going on when we push the *on* button.

And if we didn't know we needed digital media, its content will surely tell us at first glance that we absolutely do. Our devices make us smarter, more capable and more fun. Communication is a social construction and with it comes devices digitally constructed and networked to email, text, game and create immersive and augmented experiences. While some researchers debate whether technologically mediated communication is "real" or not, others like me conclude that as soon as the idea of this kind of communication entered our discourse, it becomes *real*. An old proverb I once saw hanging in a drugstore window notes, "We are changed by every person we meet." I wonder, are we also changed by each technology we meet? Does the perceptual experience of connecting with digital media change and co-shape us? I think it does. Digital media is such a ubiquitous part of our lifeworld, the world we experience and live in, that we often forget that this content and platform and device driven wave of technology is a tool. When placed in the interpretive lens of tool and artifact and instrument, we can begin to study digital media in a different way. Whether in the foregrounded or background, digital media is a kind of technology that continues to push the boundaries on production, distribution and communication. The aim of the study is to recognize that digital media technologies and the content it creates and proliferates does not lack effect. It is not invisible or neutral. It textures our world in multiple and varied ways. This acknowledgement is an important one. A technologically textured world is not broken down into positive and negative effects, but an exercise in identifying microperceptual (embodied) and macroperceptual (socio-cultural) themes.

This first section of the book,"Raw Materials," is named as such because the first three chapters share the basic provisions from which a creation is made. The introduction orients the human–technology connection created through digital media and the mediating role both technology and humans play in that connection. While many philosophical works explains the human–technology experience as a relation, this investigation used the term "connection" because this word also explains what digital media does—it fundamentally creates connection. The connections are multiple and overarching, however I use the singular word "connection" to emphasize a relationship or linkage with another. The American philosopher Don Ihde's metaphor technological texture is an insightful way to understand the interwoven strands of our relationship to technology. He notes, "Those of us who live in the industrially developed parts of the Northern Hemisphere live and move and have our beings in the midst of our technologies. We might even say that our existence is *technologically textured*" (1990, 2). This metaphor will serve as a guiding force for this book.

And Ihde is right. Technology is a texture, a feeling or surface that is knit into how we live and what we do. Even when we decide that it has all been too much and we put our technology away and "unplug" from the texture, digital media are still shaping our world. With the ubiquity of smart tools, smart phones and smart environments, the idea of "unplugging" becomes quite impossible, especially with the Internet of Things ubiquitously linking to the very fabric of our environment. This is our digitally mediated world. To be sure, the task of uncovering the human–technology connection is an unending journey, with many assumptions and caveats. Ihde's fifty years of research grounds many ideas about the human–technology connection and serves as a kind of force field for this book. His insights on technology are many, his explanations are clear and his postphenomenological lens can be easily adapted to the topic of digital media. Ihde calls today's technological experience a kind of compound vision of the human–technology-world. "This unique set of characteristics includes multiplicity of images, fragmentation into digital form (bits and bytes) and fluidity of technological artifacts (digital media) as they move in and out of presence, a total bricolage character of experience" (1990, 64). His work on image technologies and scientific instrumentation aligns with human–technology connection of digital media through a perspective that shifts from classic phenomenology to postphenomenology.

For Ihde, the world is always a negotiation between humans and their tools, their artifacts, their technology, and their devices. Technology users interpret the world through their technologies, and nowhere is this truer than in the world of digital media, where mediated interpretations are compound, fragmented, fluid, multiple and varied. The idea of artifacts, also called instruments, objects or tools, is important and central to the conversation. "In this cultural psychological perspective, all artifacts are artificial—they are not grown out of soil or shaped by the winds. Human beings create them and embed them in human activity. Artifacts are created with purposes. They are tools meant to fulfill a desire or requirement developed in a cultural group of people, who pass not only material artifacts but also their cultural meaning on to the next generations through a process of learning" (Hasse 2008). And not only are human "users" brought into the digital media experience to employ technology to do some specific task, but also as consumers, to consume the technology and all it has to offer. What desire and requirements do digital media answer?

"Raw Materials," featuring chapters 1 through 3, introduces our human–technology connection through digital media. Chapter 1 provides introductory information about our technologically textured world and lays out a basic grounding to the study. Chapter 2 contextualizes digital media within several different disciplines to better understand the texture of technology. Chapter 3 digs deeper into the philosophical framework of postphenomenology as a way of thinking differently about digital media, one that reveals the non-neutrality of our technologies. Chapters 4 through 12 provide nine digital media cases that consider the human–technology connection through a largely postphenomenological perspective. And finally, chapter 12 provides an epilogue to reconsider the technological texture of the human–technology connection.

One of the most interesting philosophical conversations about technology is its neutrality or non-neutrality. It might make sense to think that our very latest media technology is more like a neutral conduit and pass-through of information than an opportunity for effect, but this is not the case. The Canadian media philosopher Marshall McLuhan stated many years ago that the very way the content arrives to us is bundled into the medium itself, and "the 'content' of any medium is always another medium" (1994, 8). Media literacy scholar John Culkin shared that "we shape our tools, and thereafter they shape us" (1967, p. 52). Media ecologist Neil Postman (1992) explains, "The benefits and deficits of a new technology are not distributed equally"

(9) and our "tools are not integrated into the culture; they attack the culture. They bid to *become* the culture. As a consequence, tradition, social mores, myth, politics, ritual and religion have to fight for their lives" (28).

For quite a span of years newer philosophers of technology like Ihde (*Technics and Praxis*, 1978), Albert Borgmann (*Technology and the Character of Contemporary Life*, 1987), Langdon Winner (*The Whale and the Reactor: A Search for Limits in an Age of High Technology*, 1986) and Peter-Paul Verbeek (*What Things Do: Philosophical Reflections on Technology, Agency and Design*, 2005) have also consistently proved the non-neutrality of technology. But this idea about digital media has not so quickly moved into mainstream conversation. Media and behavior psychology research has shown that media exposure alters behavior (Anderson et al, 2010; Annenberg Media Exposure Group, 2010; Brown et al, 2006; Krahé et al, 2011) and face-to-face communication (Pierce, 2009; Smoreda, 2002; Turkle, 2011) but mainstream conversation squarely centers on the persuasive idea that digital media lacks significant and substantial overall effect on life. *Digital Media: human–technology Connection* focuses on work from Postphenomenology, PhilTech, Media Ecology, Media Studies and Film Studies to unpack the enhancing and reducing qualities and humanly defined values and processes of digital media to cast a holistic view that illustrates the non-neutral technological texture of our world.

Research at the turn to the twenty-first century used the phrase "new media" as a way to distinguish digital media from material "terrestrial" media like radio and television, which used analog signals, or film that used celluloid plastic for production. The ideas, principles and theories of new media are well documented and serve their purpose of explaining the transitional period of media that went from old to new (Bolter and Grusin, 1999; Manovich, 2002). I will use the term *digital media* instead of new media throughout this investigation because it emphasizes and situates the important and contemporary term *digital* (meaning non-positive and positive discontinuous bits of data) in the media process. The definition of digital media in this study is what it is, to emphasize the digital in the media process, not the newness of the technology. This holistic definition includes technology in the form of devices and processes and communicated content all rolled into one. Mainstream writing and reporting often describe digital media in terms of technologies or devices only. The term encompasses a shared texture or DNA of zeros and ones that knit so much of today's technologies together.

Digital media is an increasingly popular and universal way to mediate our connection in the natural world and to inform our socio-cultural practice. No longer just at our fingertips, technology is on and around and in our bodies and our environment, allowing us to communicate and share data about how we encounter our environment through the Internet of Things. These familiarities become the experiences of human–technology connection called situated knowledges that increase understanding about how technology co-shapes our world and the entire ecosystem that nets us (Haraway 1991). We can discover a great deal about digital media if we use continental philosophy and a variety of texts to open possibilities for a different way of thinking about the many varieties of technological experience. The book gives traction to a different kind of analysis to understand human–technology connection to digital media.

In part two of the book, called "Feeling the Weave," my inquiry moves into case studies that illustrate the nature of digital media through a variety of materials, transcribed interviews, descriptive content and academic sources. As a writer, I realize that my epistemological and ontological positionings as a scholar, a college professor who teaches digital technologies in the classroom, a professional media maker and an avid user of technologies situate my "take" on the topic of digital media and its human–technology connection. My experiences are part of the interpretations. I cannot bracket them out, but embrace them in the pursuit of understanding. The Canadian curriculum theorist and scholar Max Van Manen explains the spirit of this kind of inquiry as the reflexive practice of teacher-scholar. The German continental philosopher Hans Georg Gadamer (2000) further articulates, "A person who 'understands' a text . . . has not only projected himself understandingly toward a meaning—in an effort of understanding—but the accomplished understanding constitutes a state of new intellectual freedom. It implies the general possibility of interpreting, of seeing connection, of drawing conclusions, which constitutes being well versed in textual interpretation" (260). This book is my effort as a teacher-scholar to see and understand digital media in a different way from that which it is currently being seen. My aim is to describe experiences that uncover variations from the tiniest details to the bigger picture, to interpret the bricolage character of experiencing digital media as I see it, in its non-neutrality (Irwin, 2014a, 2014b, 2010a, 2010b).

Being in the midst of our technology means continuously living with and experiencing technological entanglement in a way that changes us. The notion of technological texture, seems a most fitting metaphor for the human–technology connection. Even the idea of networks and webs harkens back to the texture, which is a pattern and feeling that envelopes everyday life and experience. Exploring this metaphor helps clarify the Gordian knot of the human–technology relationship. The noun texture comes from the Latin root *textura* or *texera*, meaning weave. A texture is both surface and interior braiding and intertwined threads or strands that make up a kind of woven fabric. I often feel the weave as I engage with technology. Where does one part begin and another part end? Where are the edges and boundaries, the overlaps and co-shaping? Is the interlace made up of many parts or one big whole/hole?

The Russian filmmaker Sergei Eisenstein, when discussing montage editing in his book *Film Form* (1969), says, "How neutral it [raw material] remains, even though part of a planned sequence, until it is joined with another piece, when it suddenly acquires and conveys a sharper and quite different meaning than that planned for it" (10). We see this overlapping and joining when apps meet plug-ins, and patches and downloads overtly or obscurely create our montaged world. Technological texture can be an imitation or representation of something real or the result of something real, woven together to form a new whole or augmentation that extends transparency and takes the experience into one's body so that it is no longer an object of perception (Liberati, 2012).

The result of the connection between human and technology becomes the texture of the weave.

ABOUT THE PHILOSOPHY

It seems important to explore some philosophical ideas of the human–technology connection here in the introduction, but sections on postphenomenology and phenomenology in chapter 3 more fully explore these subjects. A variety of technological trajectories have been explored through the specific discipline called the Philosophy of Technology (PhilTech). Val Dusek's *Philosophy of Technology: An Introduction* (2006) and David M. Kaplan's *Readings in the Philosophy of Technology* (2009) provide depth and breadth in the understanding of major schools of thought in PhilTech. But the historical context of this important philosophic field is also important in understanding tool use, technology, artifacts and the complete conversation of digital media. Bob Scharff and Val Dusek's *Philosophy of Technology: The Technology of Condition—An Anthology, 2nd Edition* (2013) is a thorough book that historically situates early tool use life with Plato and Aristotle's ideas of *techne* and provides a historical continuum from early philosophical thinking on technology on through to present day ideas and trajectories. Scharff and Dusek's anthology explores the positivists, the pragmatists and a variety of different thinkers that build toward a philosophy that identifies and interrogates complex digital technologies. One of the most well known early Continental philosophical works to contemplate technology is the essay "Questions Concerning Technology," by German Continental philosopher Heidegger (1977). Scharff and Dusek include Heidegger's essay in their text, along with commentary from several PhilTech scholars who treat the piece as a well-known philosophical text about technology, but shift the contribution in new and divergent ways to foster new dialogues about contemporary technology (Friis and Crease, 2015), (Liberati, 2015), (Goeminne and Paredis, 2011). Texts like this one help clarify the past and present dialogue in the field as this work takes up digital media and its human–technology connection.

Early on, the anthropological perspective of technological instrumentality seemed to be most linked to human-tool use. Borgmann (1987) explains that a continuous historical thread connects today's digital media with the earliest tools and "the extension of human capacities through artifacts is as old as humankind itself. A human being is, simply, a toolmaker and a tool user" (10). Heidegger explores these ideas about the human–technology connection and describes that the best way to reveal the essence of technology is through one's questioning about or concern for technology itself. He describes that the simple relationship of toolmaker and a tool user does something to humanity's ability to live life to its full potentiality, which is different from a technology user with a technological instrument, which hides an opening that is required to be fully present. This nuance, often studied around the Greek notion of *techne*, delves into the relationship between humans and their tools. While Heidegger's work on technology is often referred to as the start of the conversation about philosophy and technology, his tool-use philosophy and ideas about metaphysics do not play a central role in the philosophical framework because of this study's turn to the perspective of postphenomenology. And while it is not essential to completely understand phenomenology to use a postphenomenological perspective, grasping an emphasis in interpretation and understanding in the search for meaning is important. Background into Heidegger's ideas also provide context to a larger conversation in the philosophy of technology and some of the ideas about technology that are still prevalent in contemporary thinking. While these insights can be helpful when considering tool use, his analysis does not go far enough to uncover essential ideas about co-shaping, co-constitution and interrelatedness of humans and technologies. Shifting from these well-established early notions to newer thinking about digital media and philosophy through current work in PhilTech loosens the weave of the texture to see what light comes through.[1]

Contemporary work in PhilTech focuses on the ideas of interrelatedness in the human–technology connection. The etymological Latin root *relates* notes a bringing back or restoration to one's being (Barnhart 2001, 1626). Might the intertwining interrelatedness with digital media be a bringing of oneself back to another? Is this human–technology relation something we specifically search for, or is connecting just part of being human? One way to explore the lived experience of human–technology connection is through phenomenology, "a philosophical style that emphasizes a certain interpretation of human experience and that, in particular, concerns perception and bodily activity" (Ihde 1990, 23). An emphasis on interpretation allows certain lifeworld proficiencies to be uncovered and explored to make a connection of the body and perception. Early phenomenology worked to define "*that* which is experienced, the *what* of experience, the 'object-correlate'" and explore the idea that "there is nothing which is present *as evidenced* unless it is present to experience" (Ihde 1983, 146). In short, there is no human act that is not directed toward some thing in the essential interrelatedness of the lifeworld.

The canon of phenomenological texts that study human–technology connection is thick but several early phenomenological thinkers, like Edmund Husserl, Martin Heidegger, Hans-Georg Gadamer and Maurice Merleau-Ponty, reveal important introductory aspects and an orientation for a postphenomenological perspective, the main philosophical practice for this book. The aim of this introduction is to explore the place of this research within the field of PhilTech, and then within philosophy, and then define several overarching concepts for background and context for postphenomenology. The origin and theory of digital media are considered in chapter 2 and the philosophy is further explored in more detail in chapter 3. But first it is important to explain working definitions of significant terms like perception, embodiment, lifeworld and interrelated ontology.

PERCEPTION

One way of viewing the way we connect as humans, in the world around us and through the media technology in that world, is by studying our perception of the sensual world. Merleau-Ponty reveals bodily connections to perception through the notion of the flesh, the body that experiences in the world, measures experience and reflects it (2000). Our body and our worlds (corporeally real and virtual) and our materiality (digitally mediated) become our environment where the intentional human–technology connection occurs. Technology users often feel joined, connected and mediated with the technology when the technological terrain is understood. But at other times the human–technology connection is disjointed or lost. Just as the weave is pulled together to become one fabric, the user comes together with technology to create one dynamic system of connections that form a texture. But this does not mean that the fabric never rips, stretches or tears. The concept of lifeworld perception and the existentials of the lived body, lived time, lived space and lived relations, provides an opening for relating and connecting.

Perception has several foundational features that capitalize on seeing, hearing, feeling, and touching as modes of experiencing the world. The senses create interplay between the figure against the ground, with every object contextualized against that environment. Each perception occurs for the perceiver from a specific embodied position and our use of digital media becomes textured and co-shaped within perception. How do the connections, the co-shaping and co-constitution for human and technology begin? As with humans, a meeting or introduction occurs. A connection is a meeting for transfer of passengers, the relation of ideas, a joining together of things. Etymologically, the idea of connection stems from the Latin notion of tying in knots and twisting or covering with a net. A connection is strong—a connection is binding with another. A texture is also a joining together, but this time the interwoven strand-like threads surround our very being in its midst. What might a texture of netting or tightly tied knots feel like as it surrounds? How might digital media bind humans together, technologies together, and human and technology together to connect and surround them? The phenomenological writer and scholar Dave Abram, in his book *The Spell of the Sensuous*, understands perception as a world of unfinished objects experienced through the senses. "Each thing, each entity that my body sees, presents some face or facet of itself to my gaze while withholding other aspects from view" (Abram 1996, 50). The idea of the body's senses playing a key part in perception is foundational for the interrelational ontology between human and technology.

What stands out first is that all human–technology relations are two-way relations. "Insofar as I use or employ a technology, I am used by and employed by that technology as well . . . A scientific instrument that did not or could not translate what it comes in contact with back into humanly understandable or perceivable range would be worthless" (Ihde 2002, 137–38).

Technology can be experienced similar to a garden. The perception of a garden is not wholly determined or caused by the flowers or wholly caused by the self. In the same way, the perception of technology is not completely textured by either the technology or the self. "Neither the perceiver or the perceived, then, is wholly passive in the event of perception" (Abram, 1996, 53). For the human–technology connection in digital media, perception occurs through multiple windows and apps, browsers and networks through the eyes, ears, skin and bodily felt senses. My gaze sees some aspects of the technology and the interface and not others. "Allowing the past and future to dissolve and merge, with a focus to the imbedded presentness, to be engaged in the present moment, is to be in the enveloping field of presence . . . a place which is vibrant and alive, which melds the time and space in a less distinct way" (Abram 1996, 204). Embedded presence, especially through embodiment, is at the forefront of the digital media connection and will play a major part in this study. Phenomenology addresses perception as part of being-in-the-world.

Husserl (1965, 1980) focuses on the structures that are essential components for understanding the human–technology connection in digital media because he specifically targets the structures connected to perception. Technology creates such a structure. Merleau-Ponty, along the same lines, notes, "For us the essential is to know precisely what the being in the world means" (1987, 6). Technology is in the world with us, as part of our world. How does "being-in-the-world" with technology make a difference in how we live? Heidegger explores the phenomenological idea of self-understanding through existence or "being-in-the-world," in a pre-ontological way in a turn toward the meaning of existence (1962). Being-in-the world brings a phenomenon into view, to be shown in the light to show itself. For the digital media user, what shows itself? The technological texture can be so transparent that showing technology for itself can be difficult. Transparency "refers to the degree to which a technology recedes into the background of user's awareness as it is used (Rosenberger and Verbeek 2014, 23). Concealing and revealing mediate the field of awareness in the human–technology connection. Studying how the body relates and connects with technology helps to reveal the technological texture.

LIFEWORLD

The *lifeworld* is the place where we work, play, live, and breathe. It is where perception begins. This term is richer in meaning than a more generic or scientific term like *world* because the emphasis of lifeworld is on the pre-theoretical place where our intentional experiences in the natural world occur. (The lifeworld is where we experience the world). One of the ways to study digital media and subjective experience is by thinking about ways technology is used in the lifeworld. Digital culture "does not imply that everyone is or sooner or later will be online and better for it, but assumes that in the ways humans and machines interact in the context of ever-increasing computerization and digitalization of society, an emerging digital culture is expressed" (Deuze 2006, 66). Exploring digital culture begins to scratch the surface for better understanding the non-neutrality of digital media. Situating the understanding of an emerging digital culture can be a prickly proposition but is squarely placed in the day-to-day lifeworld.

The philosophers Gadamer (2000), Habermas (1989), Husserl (1954, 1970), Merleau-Ponty (1962) and Schutz (1967, 1973) have all contributed philosophical work that interprets what constitutes the lifeworld (German *Lebenswelt*). In the study of digital culture the encompassing notion of lifeworld is "not an immobile one; not a merely passive and ahistorical ground, but rather lifeworld (or, better, life-worlds) as something always in motion, always in a process of sedimentation and foundation: always in a crisis" (Dorfman 2009, 300). Our technologies are not a neutral part of the lifeworld but are embedded, textured within it. The lifeworld is linked to our technologies in a way that transforms our lifeworld through our everyday perception and lived experience of it. The reflective spirit works to probe the digital media: human–technology connection that takes place in our lifeworld. Our relation within the lifeworld is rooted in the connection of subjects to objects in an intentional way, through our perception. Gadamer speaks of the lifeworld as "human life is embedded in a meaning-structure, a horizon of meaning that surrounds every act, action, articulation, or reading" (Svenaeus 2001, 154). His two words for lived experience, *erlebnis* and *erfahrung,* distinguish between immediate personal experience and a collective social interaction and historical experience in tradition and community. "Erlebnis is something you have. Erfahrung is something you undergo as an open and ongoing experience" (Ghosh et al. 2007, 21). But erlebnis is also the specific and personal in light of the whole. One cannot be fully distinguished from the other. This distinction of lifeworld further identifies the complexity of living in the world, which opens further understanding of technological texture.

EMBODIMENT

The idea of the body in technology begins with the notion of taking thinking down into the body (Levin 1985). When thinking goes to the bodily felt experience, human–technology connection begins. I cannot bracket out my body in digital media: I need to listen to my body, because mediation cannot occur without the body's knowledge. But that does not mean that I am always perceiving the experience. Embodiment is rooted in perception and technological experience changes as a user's body adapts or does not adapt to digital media use. Virtual reality sickness is one type of motion sickness that makes a user uncomfortable while engaging in technologically mediated activities. The perceptual focus becomes the back and forth of disembodied and embodied perception.

> Ontology that is founded on a corporeal schema which roots the human body, as a local opening and clearing, in the multi-dimensional field of Being, for it articulates the embodiment-character of our responsiveness and elicits its potential for development on the basis of our initial, most primordial sense-of-being-in-the-world." (Lakoff and Johnson 1999, 62)

Understanding the idea of the body in terms of flesh can help to explore the idea of habit and how the user's body does things with unconscious awareness. Merleau-Ponty's concept "well in hand" (2000, 143) describes the experience of complete accessibility and capability through the flesh and the hands. For many digital media users, the hands are a central component in the connection. When a user is habitual with digital media he or she can see what is in reach, and the spaces on the screen and the keypad do not stand out but are incorporated into the body schema. The device is "well in hand." Texting without thinking about it, moving around a gaming world and tapping on technology to alter programs and information flow become enmeshed within the texture of the technology as "habit expresses our power of dilating our being-in-the-world" (143). This is embodied motility, based on habit and bodily intuition. The mind also is engaged in the process in a profound way. Many kinds of digital media require patterns that are built right into the operation to be highly individualized and habit forming. The user is enveloping the device into his or her bodily space and the body becomes the mediator of the lifeworld. Essence and existence meet here to form a human–technology connection.

Digital media is designed to incorporate habit-forming body knowledge. Inhabiting digital media means the user is engaging in a hierarchical kind of learning that results in comprehension from practice. In the discipline of cognitive psychology, researchers have focused on learning through hierarchical control as part of the practicing process. When someone learns to text, game or work through an app, they engage in a "hierarchy of control built up from elementary units to high level constituents" (Johnson-Laird 1988, 210). This is called "chunking," when the brain memorizes information and improves learning through practice. There is far too much information impinging on the senses for it all to be processed in full. "A selective filter therefore operates to determine which information shall pass through a 'channel' of limited capacity for further processing" that occurs with practice. The knowledge base moves from a fragmented state of view about a process to an appreciation of the total picture of the process (147–48). Practice seems to accomplish the task of recalling several pieces of information at once to increase the ability to negotiate the device and information repeatedly. Embodiment is the kind of body knowledge that becomes central to human–technology connection. This leads us to our last concept: interrelated ontology.

INTERRELATED ONTOLOGY

A phenomenological exploration begins with the idea of existence and our relation to the things around us called ontology. The term is explained as the "whatness" or "essence" of an encounter—in this case, our encounter with digital media. The origin of *ontology* comes from the two Greek words *on,* which means "being," and *logia,* which means "study." Ontology is a branch of metaphysics that studies the existence of Being. Ontology, in a sense, is the vision of the world we have through relating in the world. How can we relate and connect in a different way to experience our human–technology connection because of our digital media (non-human) technology?

When digital media is ontologically designed, it means the software designers have made the software easily relatable on a human level. The phenomenological writer David Michael Levin explains, "Ontology does not happen by itself. Ontology is a work of thought, and therefore it must be referred back to (or correlated with) the being who is thinking. As an undertaking of human beings, ontology manifests the character of the human being who is always already in relatedness-to-Being" (Levin 1985, 23). Heidegger (1962) calls ontology the phenomenology of being. He emphasizes our need to understand the importance of the ontological priority or existence "of" and relating "to" being. Ihde pushes the notion further to explain an interrelational ontology often used in describing the human–technology connection. "This style of ontology carries with it a number of implications, including that there is a co-constitution of humans and their technologies" (2009, 22). Until recently, there has been no specific work on ontology of technology, but co-shaping with technology is becoming a more studied concept in social sciences (Lawson 2008). Interrelational ontology explores how we are ontologically related to an environment in which the user and the technology are both transformed in the co-constitution.

Technological texture is the exploration of the interrelationality, the inbetweenness, the hyphen of the human–technology connection. How are we experiencing the world with-in our digital media? Is the surface a real texture one is touching, or is technological texture only an imitation or representation of a real felt experience? How does the experience change the nature of digital media use? "The objects (nonhumans) in such interactions modify the humans, the subjects are nonneutrally and noninnocently invariant, but the counterpart modifications are not those of immediate, real-time modifications . . . If we 'dance' with the non-humans, the steps that occur are often different from and often out of tune with the music played" (Ihde 2000, 100).

One of the most foundational concepts for this kind of dance is studying how the human–technology connection might be different, or "out of tune," from a human-to-human connection. Sometimes access to and use of technology seems like a matter of life and death. I have heard people say, I could not "exist" without my cell phone or how did you "live" without this or that technology. Technology may be required for medical reasons, or desired for many reasons. As bodies in relation to technology, our perception and environment are mediated by the technology. This is interrelational ontology at work to con-constitute and co-shape the human–technology connection.

WEAVING IT TOGETHER

This book claims that the digital media human–technology connection can be understood in new ways through a different kind of philosophical study. This book aims to do just that by bringing meaning to the digital media experience by studying contemporary lifeworld experiences of digital media. How might understanding our technology's effect on us change how we live in the world? Gadamer (2000) emphasizes the idea of understanding as a way to know being-in-the-world. Reading about technological experience and studying transcribed texts about digital media specifically can increase understanding. Philosophy of communication, media ecology, semiotics, critical theory and critical-cultural studies, and a host of other methods and disciplines also explore the human–technology connection of digital media. Each brings new information to the conversation. But the conversation is ever evolving, like the digital media it studies. It seems to make sense to continue to try different approaches to reach an even richer understanding of the technological texture, specifically, to further identify the non-neutrality of the connection.

We continue in the spirit of "Raw Materials" to explore the many ideas that encompass the technological and theoretical concept of digital media. Chapter 2 works to contextualize this concept within the study and describes convergence, digital design, mediation and mediatization along with media theories that further frame the human–technology connection. The aim is to always keep in mind that digital media, like all technology, does not lack effect.

NOTE

1. The aim here is to give context to postphenomenology by providing some related aspects of phenomenology, not to specifically identify and explore Martin Heidegger's ideas about technology, his questions concerning technology or his metaphysics. The focus is on postphenomenology as developed by Don Ihde and others (e.g., Ihde 1990), that deal more with the practical description of everyday relations to technology than being, *Dasein*, and essences.

Chapter Two

Describing Digital Media

Digital media was loosely explained at the beginning of chapter 1 as an overarching name for a variety of technology and content associated with media. The Raw Materials section continues in chapter 2 with specific ideas to ground digital media as more than the comingling of content and digital tools. Digital media is the contemporary name for digital content and digital devices like smart phones, tablets, computers, televisions, players, watches, gaming consoles and even billboards. Fleshing out a more concrete understanding of digital media as it relates to prior analog technology will better illustrate its role and unique properties as a mediator of human experience and practice.

Before media content was available in digital form, it was created through an analog process. This change in technology created an environment that altered how content could be created and shared. Analog, simply put, is an electronic process with a varying signal that conveys another signal. Analog signals use the property of a medium to convey that signal's information. Traditional terrestrial media, like radio and television, were born of analog technology. An analog signal picks up sound waves and broadcasts them as radio programming through a fluctuation in air pressure to send the programming to the transducer (the radio tower). Analog signals pick up visual and audio waves to create a signal read by an antenna. Continuous variations of the signal produce different aspects of the signal to create different sounds and colors. Electronic media were analog until computers and networking made them digital. The analog process is different from the digital process because analog signals broadcast information, which is translated to electric pulses of different sizes (amplitude) and carried on waves to a receiver (tower). Digital signals translate the information in a binary code of 0s and 1s so the information is received to its final destination without any loss of information through two separate amplitudes. Digital signals can be exactly replicated over and over again. When data is transferred through an analog system noise and signal degradation can occur. Digital transfer offers endless and multiple uses. This difference inexplicably altered the creation and reception for communication and media legacy to create the contemporary digital media environment (Couch, 2012).

The content for digital media is stored in numerical form. "The fundamental quality of digital media is that they are driven by minute, discrete electrical impulses, commonly characterized as 'on' and 'off'" (Cubitt 2009, 23). The digital form allows the digital signal to move quickly, especially with increasingly optimized bandwidth, expanded wireless networks (wifi) and more efficient compression, to inhabit any platform or device that will have it. All of this requires digital infrastructure like high-speed broadband, fast connectivity, secure data carrying, regulation that encompasses local, regional, national and international governing bodies, and digital media devices that work within this paradigm. The term *digital media* describes a mix of digital components—computers, software and networks—combined with media characteristics like the information and content created using digital tools. The content is disseminated back through digital devices and then experienced within a specific use context. Another way to describe digital media is to consider the three "parts that distinguish them from previous media forms, content (discourse), channel and application" (Aarseth 2003, 319). All of these ideas encompass what I call *the whole of digital media* and meaning of the term throughout this book.

One of the reasons digital media is what it is comes through convergence, or the interconnection of content, digital media devices, networks and information. The idea of digital convergence, moving from many devices with different contexts and content to one device with many kinds of content and use contexts, increases the focus and attention and need/convenience of relying on one digital device. The separation of media content from specific digital media devices (TV programming watched on a TV) has created a proliferation of ways to use just one device. For instance, a phone can be used to talk, text, gain directions, watch television, listen to portable sound, play games and collect self tracking data to name only a few uses. And then the actual use creates information for surveillance and big data collection. Convergence requires cooperation of many industries and its result alters the way we communicate, relate, think and act just because it is available. Understanding convergence opens the weave a bit, to see through and into the digital media design. Important for this study is an understanding of human–technology connection through digital media. The digital side and the media side create interactivity, so digital media works like a feedback loop (Kaptelinin and Nardi, 2006). Unlike mass media, where one can communicate with many, digital media allows one person to communicate with many and also allows many to communicate with many, and one to communicate with one, in many different ways. The whole of it is called *digital media*. For this very reason, digital media is not a neutral technology. To be sure, many studies explore wide-ranging ideas about digital media and the associated term, new media. See Dewdney and Ride (2014), Jenkins et al (2013), Kovarkik (2011), Creeber and Martin (2009), Mark B.N. Hansen (2006), Liestol et al (2003) and Rodowick (2001) for depth and breadth on a historical, cultural, artistic, and aesthetic understanding of digital media.

This chapter serves to orient, explain and explore many aspects of digital media. Through a combination of unpacking and digging deeper, the warp and weft of the weave of the technological texture can be seen. The metaphorical comparison between weaving and contemporary digital media is designed to set certain ideas against others to illuminate the bifractured and bricolaged character of the texture. In the craft of weaving, the warp is the tightly stretched lengthwise threaded core of a fabric and the weft is woven between the warp threads to create varied patterns. The filler thread (weft) fills the design while the warp defines the design. Warp threads are the stronger, coarser threads that provide the core of the textile, and they are placed first on the loom. Each part contributes to the design, but the complete pattern is all that an observer sees on first glance. It takes further study, even possibly turning the textile over to see the underside, to reveal the whole of the weave. Digital media too has a warp and a weft. Are the humans the warp, the stronger side that provides the core of the weave? Or the weft, the filler thread that fills the design? In the same way, description, clarification and case studies contribute to better understanding the entire technological texture of the lifeworld.

ABILITIES

Digital media has some unique qualities and abilities from its analog media predecessors, namely spreadability and usability. *Spreadability* means that digital media is technically and culturally spread through the participatory culture of sharing on social media and other online networks (Jenkins et al, 2013). This ability to spread quickly "refers to the technical resources that make it easier to circulate some kinds of content more than others, the economic structures that support or restrict circulation, the attributes of a media text that might appeal to a community's motivation for sharing material, and the social networks that link people through the exchange of meaningful bytes" (Jenkins et al 2013, 4). In short, digital media are spreadable because they are mediated by a computer and kept "new" because they are digital. This environment proliferates a constant increase in technologies (Gladwell, 2000; Kurzweil, 2005).

The technology "containing" the digital media is called the *device*. Devices are the smart phones, gaming consoles, tablets, music players of all varieties, and wearable technologies like glasses, watches, and clothing. Devices also ubiquitously foster biotechnology and automation. Connection is the key to the process of digital media. The experience is a relation and an entanglement, but the fundamental connection to the device makes spreadability happen for digital media.

One other unique characteristic is digital media's *usability*. Great care has been taken to create devices that easily connect and are universally used and adopted no matter what platform, proprietary operating system or service that is used. I happened upon a website one day that posed this question right in the middle of the homepage, "What type of user are you?" The website was designed to identify right off which portion of the website might be most marketed to me. Was I a new or repeat customer, an employee or a consultant? How was I going to "use" their site? This experience stuck with me as a question central to the human in the human–technology connection. Digital media spends a lot of time telling us what we should be using, the shortcuts we can take and how to more quickly or more efficiently use our digital media.

The connection between a user and his or her digital media is a technologically textured one that is often ubiquitous, sometimes ambiguous and seemingly complex. The human user loves technology when it is new and/or seems to fulfill many of our communication and living needs and quite a few other cultural desires like status and companionship. While the term *user* is problematic in many ways, it explains an important side of the human–technology connection and recognizes that humans employ the technology for specific uses and the technology needs a user at some intersection of the process to make it work. This description of the human in the human–technology connection will be employed throughout this book to name the connecting role. The term *user* in the context of digital media means both humans who use digital media and those who are potential future users. This use context helps describe the human–technology connection because even the very creation of a technology means the world is shaped differently by it being there for use, which further explains the ways it fundamentally changes the lifeworld. Explains media ecologist Neil Postman, "[I]t is a mistake to suppose that any technological innovation has a one-sided effect. Every technology is both a burden and a blessing; not either-or, but this-and-that" (1992, 4–5). He pushes the notion further by noting, "To a man with a pencil, everything looks like a list. To a man with a camera, everything looks like an image" (p. 16). The non-neutrality of digital media will play a major role in this book.

In the design process, digital engineers are constantly balancing the usability and ergonomic styling of a product. They study how humans will use the technology and how their experience might vary based on certain design features. "[P]roducts are not merely tools: they can be seen as *living objects* with which *people* have *relationships*" (Jordan 2002, 7). The Apple Watch, tethered to the iPhone, is marketed as one of the most intimate technologies on the market. Talking, touching and drawing on the device all create digital chain reactions that make multiple tasks and activities happen. With a high level of internal customization, notifications like "glances" and varied audio buzzers coupled with a sophisticated interface try to integrate functionality seamlessly into daily life. Usability seems central to the human–technology connection.

American courts have ruled that one's mobile device is a private technology with personal information connected to it or stored in it. Digital media has become so individualized that fingerprints have become a way to access a personal device. Users want their device flexible, useful, and individual. Personal privacy has become an increased focus for research and development. Historically, digital designers ignored the human factors in the design process. Through time the process shifted slightly, and after the engineered product was complete, human factor designers were brought in to give the product an inter-face lift of sorts. At first this shift brought with it superficial nods to human factors in the look and even the feel of the design, but no real engineered differences occurred in the human–technology connection. However, today's digital tools are clearly part of the new philosophy for creating digital media devices that allow the human factors to be inseparable from the design process (Jordan 2002, 2). "Our human–technology relation has been facilitated by design experts who approach the creation of digital media devices by first, understanding people holistically, linking products benefits to product properties, and developing methods and metrics for assessing product pleasurability" (8). This kind of strategic thinking facilitates the in-between of the technological texture.

Designers and engineers have moved beyond labels like "user-friendly" and "user-centered" to focus on human factors, which deal with how we connect with technology. Engineering psychologists have joined designers to create visualizations, haptic sensors and a variety of displays for virtual and 3D rendering. Heads-up displays (HUD) that do not require users to look away from their horizon have long been used by the aviation industry and the military, and are now increasingly available for consumers in vehicles, games and a variety of sports goggles. Researchers in the field of virtual reality are creating technology that replaces a user's field of vision in increasingly seamless ways. The sensations of feeling like you are "inside" an activity, be it a game or a history lesson, simulate a new kind of presence that melds the real world with the simulated one. Either way, the experience is lived. This work on presence is a fundamental embodied shift for human–technology connection.

Also paired with the notion of the user, is the consideration of the consumer. While digital media consumption is often measured through an economic lens, consumer culture theory can identify consumer behavior through a social and cultural lens. Participation in the digital media lifeworld involves the initial purchase of the device and then access to information, technological infrastructure, web access, capacity/space, speed of connection, upgrades, downloads, and more.

> The shift toward interactive media, peer-to-peer forms of media communication, and many-to-many forms of distribution relate to types of participation that are more bottom-up and driven by the "user" or "consumer" of media (Buckingham, 2008).

In short, it is impossible to use digital media apart from being a consumer. This also then, situates digital media in a political light. Access, infrastructure, and connectivity divide and conquer. Materiality issues contribute to digital ontologies that shape political priorities and cultural protocols (E. Gabriella Coleman, 2010).

Understanding the basic definition of digital media and its increasingly virtual presence in everyday life informs the human–technology connection in the lifeworld. It seems to make sense to mention just a bit more about the technological devices, also called the equipment, the artifacts, the tools and/or the instrument, that make digital media what it is. Chris Nagel, in his essay, "Empathy, Mediation, Media," describes his experience using media technology:

> Experience of media is most directly an experience of technological equipment. My intentional directedness in an experience of media always involves dealing with this equipment, regardless of what further aversion of attention is taking place. (2010, 353)

Technological directedness is one way to describe the human–technology connection. Devices are used *for* something or *to do* something. Nagel explains that users become organized based on the requirements of the equipment they use. Designers are constantly at work trying to make technology (devices) user friendly, transparent to "hide" the technological "guts" in our lifeworld so they can be smaller, lighter and seemingly more integrated into our lifeworld.

Technology can also provide different kinds of connections between humans, between humans and devices, and between devices. Terms like *mediation, mediatization, medialization, mediatic* and *remediation* explore the nuances in understanding digital media's non-neutrality.

Kember and Zylinska, in *Life After New Media: Mediation as a Vital Process*, state mediation involves recognizing "our entanglement with media on a social cultural as well as a biological level" (2012, 1). The deeper meaning of media provides guidance for the richer connection humans might have with their technologies. Marshall McLuhan's well-known, 1964 proclamation that the medium is the message illustrates the need to critique the technological part of digital media. What is the message that digital media carries with it? If "the wheel is an extension of the foot; the book is an extension of the eye; clothing is the extension of the skin; and, very importantly, electric circuitry is an extension of the central nervous system" (xi), then of what is digital media an extension? Does the medium, the form that the message takes, knit itself into the message? Yes, as a networking of all senses in simultaneity of all things. The medium influences how digital media is perceived and received and in essential ways, becomes more important than the message of the content.

> The content of the medium varies from day to day, but the medium itself creates the environment. Once one understands that a medium is a singular grammar of its own and a combination of effects within the environment it has created, one looks at each medium in a different way, whether it be a computer, an automobile, a telephone, or a fast food chain. (McLuhan et al 1980, p. xii)

Networked and simultaneous digital media changes the human–technology connection. The grammar and effects it creates are biased in nature and vary in affect. In *Mediatization or Mediation: Alternative Understandings of the Emergent Space of Digital Storytelling* the British scholar Nick Couldry explains the way media assembles in our lifeworld. *Mediation* is the outcome of "flows of production, circulation, interpretation, and recirculation" (383). *Mediatization* is a description of the converged social, cultural and technological processes that create a format suitable for usability, spreadability and digital sharing. His thoughts on synthesizing mediation and mediatization inform the process of meaning making within the media studies landscape. The digitization process of the created media content affects the final output to codify the "logics of use and social expectations" of digital media (Couldry 2008, 383). Norm Friesen and Theo Hugg (2009) explore the historical implications of terms such as mediatization, medialization, mediation and modality and historically connect the terms with Aristotle's time up through Marshall McLuhan's important media ecology work and on to the present day. For Friesen and Hugg, the notion of "mediatic turn," digital media's interpretation of every aspect of everyday activity, is helpful to further reveal human–technology connection (64). As they discover, the mediatic turn is a "cultural, or indeed, an epistemological and existential condition or exigency." The idea that the mediatic turn proceeds the *semiotic* (dealing with symbolic understanding) and *discursive* (use of conversation) turns in history makes sense, guided by their assertion that the mediatic turn notes the middle or mediator in the process and not the total givenness of everything in the lifeworld. The idea of remediation rounds out an understanding of the different forms and rhythms media takes. "The desire for immediacy leads to a process of appropriation and critique by which digital media reshape or 'remediate' one another and their analog predecessors such as film, television, and photography" (Bolter and Grusin, 1996, 2). This reuse of content fosters the history of media technology as an embedded participant in the making. "All mediation is remediation because each act of mediation depends upon other acts of mediation. Media are continually commenting upon, reproducing and replacing each other, and this process is integral to media. Media need each other in order to be media in the first place. Once we notice this process in contemporary society we can identify similar processes throughout the last several hundred years of Western visual representation" (17).

This provocative interweaving of media concepts allows for the gradually changing form of digital media to be seen more clearly, and in a different light. As Anna Munster, in her text *An Aesthesia of Networks* (2013) explains, "In this atmosphere that is generalized communicability of networks—a complete media environment in which mediation between humans and non humans has retreated into the ambient background, leaving only mediality—there are two possibilities for the human. Either we make panicked attempts to reinstate the disjunction between things and us, or we try for relational reinvention" (189). Either way the relation changes the connection between human and technology.

Still other experiences are explained as a "love" relationship that becomes more than a user-and-tool relationship. Van Den Eede (2013) takes on many of the ambiguities of philosophy of technology and media studies and fine-tunes the human-media connection for a clearer picture of mediated experience. His schematic overview (441) and concrete examples provide a jumping-off point for some of the ways of pulling critical components from several different schools of thought and unifying them for philosophically thinking about the human-media connection. He clarifies,

> Framed as love relationships, then, our bonds with technologies and media abruptly appear as projects, undertaken through time and space. As soon as we view technology through the 'love' prism, it can no longer be seen as a 'force,' for we engage into a more or less conscious involvement with it. But it can also not be delineated as just a 'thing' for within the bond it becomes difficult to tell where 'we' end and where the medium or technology begins. (2013, 25)

Digital media seems firmly embedded in the love relationship between human and technology, where the texture begins to be knit within the hyphen of the human–technology entanglement. Contemporary digital media creates a relationship that pushes the boundary between where the human ends and the medium/device begins. They are continually intertwining in the lifeworld continuum.

CONNECTION

The word *technology* comes from the Greek word *technologia*, meaning the systematic treatment of an art. The Greek *techno* means using and relating with technology in society, including our environment, our lifeworld. The suffix *–ology* describes the branch of knowledge dealing with a notion or sense of things. The technology is somehow and in some way perceptually positioned or located between the perceived and the perception with the between-ness in the connection of the seer and the seen, with the technology changing one's relation *to* the technology. So what is the perceptual connection between humans and technology in the lifeworld of digital media? One of the most basic understandings involves humans, the lifeworld, and the instruments, artifacts and technologies (tools of old) that stand "between." Ihde clarifies that when technology is not part of the lifeworld, humans relate in an "I-world" perception. There is no intentional perception in any specific direction. This model helps to situate the place of the Self in relation to the Other in the lifeworld in a non-directional way. There is no intermediary, no technology to stand between. The metaphorical model shifts to the instrumental human–technology-world relationship when technology becomes part of the equation. This shift means that digital media has moved to a mediating position and has taken the place of "relation" in Ihde's model.

The Dutch philosopher Peter Paul Verbeek's study *What Things Do: Philosophical Reflections on Technology, Agency, and Design* (2005a) explores the change that occurs when technology becomes part of the perceptual equation. The "intentional relation between human beings and the world is thus, as it were, extended or stretched out through artifacts" (124). The focus on technology as "other" grounds the perceptual tensions between humans and instruments, as a textual artifact. Intentionality is one specific perceptual act of Being, situated in an objective way in relation to the Other. In the case of human–technology connection, the Other would be the digital media, that artifact that extends or stretches our perceptions. When perception changes and focuses upon the instrument, instrumental mediation causes a shift in the horizon, bringing one element of the lifeworld into perceptual focus before another element. Examining Ihde's model of perception illustrates the shift in this relational process to I-technology-world relations and shows an essential difference between embodiment and hermeneutic relations to illuminate the body's microperception. The connection between the technology and my relation to it lies within the lifeworld. In the human–technology-world of digital media, the human is now focused on the technological aspects of the digital media. I focus on the technology in an intentional way—with my perceptual focus on the technology as it is situated within the world. I am on one side of the lifeworld and I perceive technology on the other side. The technology is situated between the world and me. My corporeality, my body schema, is reimagined as I-technology-world connections are explored through technology.

The next shift in this model adds parentheses to the phrase, focusing on embodiment relationship and the kind of relationship that takes the viewer's body into his or her experience by perceiving through the body and the senses without noticing the technology. The "I-technology (world)" model puts emphasis on the embodied relationship between the human and the technology. This shift might be explained through the idea of playing games, talking on a mobile phone with a loved one, or wearing digital media like an iPod or a watch. I embody the technology through something I wear and view the world through this perception to create a symbiotic relationship—the technology becomes perceptually transparent. In this way the user is viewing the world through the technology, as one might view the world through the lenses of a pair of glasses, without really noticing it. I am on the "same side" as the technology in a bodily way.

The next shift is a hermeneutic one. The model moves the parenthesis to gather the technology and world into the same perceptual sweep. In "I (technology-world)" relations, the hermeneutic relations involve reading the symbols and language of the technology. This is a different kind of relation, which connects the instrument with the referent to "read" the text. This kind of relation calls on our linguistic abilities and our capacities to read through the technology, but also toward the technology in an interpreting way. I am now again on one side and technology is on the other side, but I am reading and communicating with the technology hermeneutically in the in-betweenness of the relation. Learning icons, software symbols, and the basic layout of digital tools foster hermeneutic perceptions.

The "I-technology (world)" model explains the way technologies receive multiple kinds of attentions. Technology like a heater can remain in the background, while a mobile phone or music technology can be foregrounded and focused upon. The user experiences a relation "to or with" the device in an alterity way when the technology becomes the "other" (Ihde 1990, 97). Alterity relations may be noted to emerge in a wide range of computer technologies that, while failing quite strongly to mimic bodily incarnations, nevertheless display a quasi-otherness within the limits of linguistics and, more particularly, of logical behaviors. Quasi-otherness shows that humans rely on technologies that emerge as focal entities which receive varied and multiple attentions and perceptions that can be "foregrounded and backgrounded at perceptual will 'as' other *to* which I relate," (Ihde 1990, 107). Digital media devices evoke quasi-love and quasi-hate emotions because of the notion of alterity relations. This quasi-otherness influences a user's relation to technology. In any of these ways of relating, the user inhabits the technology in the world as being-in-the-world-with-technology.

This corpus of metaphorical models shows the way we, as users, experience our devices through the different treatment of the technology and through different forms of instrumental mediation, from relational to transparent. Human connection with digital media works the same way. In the world of human–technology connection, the digital media relationship can center on one specific kind of relation or shift through each metaphorical model in any order, from background to embodiment to hermeneutic to alterity relations.

Lastly, it seems important to make one more conceptual shift that moves deeper into co-shaping and co-constitution. When a human–technology relation moves from general to authentic relating, similar to the love relationship mentioned earlier, the relation becomes one of dwelling with the Technological Other (Irwin 2005, 466). This being-in-the-world-with-technology is often experienced when art, craft and technology meld together in creative and artistic *techne*. Ontologically, users have the view of the world that does not separate the human from the technology. To even press this idea of technologically mediated intentionality even further, consider Verbeek's Cyborg Intentionality, which brings both hybrid intentionality and composite intentionality into consideration (Verbeek 2009, 390). This entanglement moves beyond technological mediation, because the intentionality is partially constituted by the technology, which gives the technology agency in the process. Virtual reality and 3D immersive technologies are examples of Cyborg Intentionality. Notions of transhuman and post-human experience become a new reality for study. Notes Verbeek, "Mediation does not simply take place between a subject and an object, but rather co-shapes subjectivity and objectivity . . . Humans and the world they experience are the *products* of technological mediation, and not just the poles between which the mediation plays itself out," (2005a, 130). Surely we, as users and consumers, are products of our digital media.

Chapter 2 has defined overarching terms used throughout the book to orient key ideas for studying digital media and human–technology connection. The final chapter in the first section of the book, Raw Materials, explores the deeper philosophical underpinnings of postphenomenology to further study the technological texture.

Chapter Three

Digging Deeper through Phenomenology

With an orientation of technological texture in chapter 1 and a grounding of digital media and the human–technology connection in chapter 2 behind us, we can move on to a basic understanding of the philosophical framework this book uses to open the weave of the texture, namely the orientation of post-phenomenology, which encompasses a theory of technological mediation (Verbeek, 2015). This chapter concentrates on building a case for uncovering the human–technology connection in a different and unique way.

Studying and understanding the perspective called *the whole of digital media* begins with a first act of communication. From the moment of that first cry at birth, communicative praxis occurs (Schrag, 1988). Babies learn quickly which processes yield what results. A variety of infant technologies are available to aid in those first months of life. The majority of these are not medical, but convenience and amusement oriented. Babies learn basic tools as they grow into toddlers. They play with toys in both intended and unintended ways. Media is integrated into young children's lives through mobile phone games; iPad activities and visual programming that are often, but not always, specifically designed for them. The communication and the technology begin to interweave and the texture becomes embedded. And eventually, for some, technology becomes a voice, and texting becomes the preferred way to communicate. In today's digital technology rich world, much of our communication is experienced through digital media, but the forms and patterns of digital media are not absolute. They vary in possibility, they vary in use and they vary in perception. While an engineer or designer may create digital media to do one thing or another, this does not mean that the specific technology's usage lacks other, equally captivating, important and exact uses. Material technologies like digital media play a mediating role in the lifeworld, but exactly how is digital media non-neutral? Which experiences are amplified and which are reduced or backgrounded when using this technology? Postphenomenology fits nicely as an interpretive framework to investigate these questions.

The postphenomenological perspective owes its beginnings to the phenomenological tradition but the framework takes a pragmatist turn when the methodology shifts to a practical, material examination based on case study analysis. Pragmatism is linked with postphenomenology because the philosophical movement featuring scholars like William James, John Dewey, C. S. Peirce and Richard Rorty aims to clarify the meaning of abstract ideas and the study of the consequence of things more than their origin. This path also separates postphenomenology from phenomenology. Both postphenomenology and pragmatism move away from dystopian ideas about technologies to identify the non-neutrality of technology that encompasses both positive and negative effects. Concerning the study of digital media, the two link in the areas of communication and media experience (Hickman, 2008). Additionally, the idea of functionalism illustrated in postphenomenological case studies further identifies "a turn, instead, toward pragmatism's orientation, most marked in John Dewey's work, toward concrete cultural (historical, social and political) events as occasions of the interaction of environments and organisms" and technologies as practices that identifies the non neutrality of technology (Langsdorf 2015, 46). Postphenomenologists

> study technology in terms of the relations between human beings and technological artifacts, focusing on the various ways in which technologies help to shape relations between human beings and the world. They do not approach technologies as merely functional and instrumental objects, but as mediators of human experiences and practices. And second, they combine philosophical analysis with empirical investigation. Rather than "applying" philosophical theories to technologies, the post-phenomenological approach takes actual technologies and technological developments as a starting point for philosophical analysis. Its philosophy of technology is in a sense a philosophy 'from' technology. (Verbeek 2015, 190)

Robert Rosenberger and Peter-Paul Verbeek, in their book *Postphenomenological Investigations: Essays on human–technology Relations,* explain, "Postphenomenologists study the relationships that develop between users and technologies" (2015, 1).[1] Their work serves as a clear explanation of the postphenomenological framework and contextualizes the empirically focused framework against prior work on the topic in a holistic and compelling way. They make clear that all postphenomenological studies share three main components: understanding the roles of technologies in the human–technology-world experience, experimental cases to reflect this experience and an analysis of the co-shaping or co-constitution of human and technology to create the human–technology connection. As explained in chapters 1 and 2, the ambition of this work is to investigate digital media technologies and the content that moves through it to point out processes and structures. How might the connection between users and digital media be understood differently from the neutral and branded way they have been presented to us in the lifeworld?

My dissertation advisor in graduate school wrote the words *dig deeper* in the margin of specific paragraphs in many of the drafts of my hermeneutic-phenomenological investigation manuscript to remind me to delve into an idea beyond the surface experiences I might be able to name right off the top of my head. I hope that this chapter on digging deeper through postphenomenology does more than scratch the surface. My intention is to bring concepts that introduce the basic tenets of postphenomenology to the digital media tapestry to further explore the human–technology connection. The idea is to show that digital media's human–technology connection is different from other technologies and different from its contemporary perception. The sizeable grouping of philosophical texts referred to here might not seem like a likely gateway for studying digital media, but this rich collection provides language and a philosophical foundation for exploring human–technology connection in our lifeworld.

As a postphenomenologist, I wish to uncover how digital media shape choices, actions and experiences in the world and how digital media change our day-to-day lifeworld experience. Technological transformations through digital technology are non-neutral because technologies are not neutral. For instance, some talents are useful in a lifeworld filled with technology, and some are not. Over time we become conditioned by our technologies in non-neutral ways. Postman argues "the medium itself 'contains an ideological bias'" (1979, 16). He reasons that:

1. because of the symbolic forms in which information is encoded, different media have different *intellectual* and *emotional* biases;
2. because of the accessibility and speed of their information, different media have different *political* biases;
3. because of their physical form, different media have different *sensory* biases;
4. because of the conditions in which we attend to them, different media have different *social* biases;
5. because of their technical and economic structure, different media have different *content* biases. (Postman 1979, 193)

Postman's belief is clearly echoed in contemporary digital media. Human–technology connection is altered in non-neutral intellectual-, emotional-, political-, sense-, social-, and content-based ways on a daily basis. Postman further explains, "The printing press, the computer, and television are not therefore simply machines which convey information. They are metaphors through which we conceptualize reality in one way or another. They will classify the world for us, sequence it, frame it, enlarge it, reduce it, and argue a case for what it is like. Through these media metaphors, we do not see the world as it is. We see it as our coding systems are. Such is the power of the form of information" (Postman 1979, 39). Postphenomenology gets at the artifact that is at the heart of the connection.

As Rosenberger succinctly explains, "In the postphenomenological perspective, a technology is conceived as an artifact which comes between a user and the world, transforming the relationship between them. A technology plays a mediating role; it transforms a user's abilities to perceive or act upon the world" (2012, 83).

This book seeks to nudge the list of digital media-related disciplines and fields a bit more to incorporate the postphenomenological perspective and framework into their research agendas. Technological and scientific disciplines like computer engineering, application developers, information design, architecture and digital innovators and entrepreneurs help transform our experiences, perceptions and interpretations with the technology they create. We make decisions based on the content provided through that technology, which transforms our being-in-the-world. An increase of research and development becomes both necessary and desirable. Applied phenomenology by way of postphenomenology provides a strong framework for the undertaking. This investigation retains and emphasizes the use of phenomenology as an analytic tool, and then takes an experimental turn to analyze the human–technology interplay through observation. "This means that postphenomenological claims are never about the absolute foundations of reality or knowledge, and never about the "essence" of an object of study. Instead, postphenomenological claims are posed from an embodied and situated perspective, refer to practical problems, and are empirically oriented" (Rosenberger and Verbeek 2015, 1).

The examination of phenomenology in this chapter is designed to identify some of the conditions of the study and provide a background or backdrop for understanding postphenomenology as it is situated in this work. This chapter also provides context for important ideas that relate to human–technology connection and the cases that are explored in the second section of the book. While the aim is to empirically explore case studies through postphenomenology, the intent is also to reflect on the human–technology connection as it is practiced. The reflection is embedded in the deep descriptions guided by phenomenological writing.

PHENOMENOLOGY

This book cannot do justice to the whole body of history of phenomenology and it isn't trying to. Reading an introductory text is a good way to understand central insights of the associated philosophical movement that historically comes before the philosophy of postphenomenology, namely phenomenology. Robert Sokolowski's *Introduction to Phenomenology* (1999), Dermot Moran's *Introduction to Phenomenology* (2000) and Shaun Gallagher's *Phenomenology* (2012) are good starting points for understanding the complete historical timeline, spectrum of ideas and seminal philosophers for this movement in Continental Philosophy. This book uses philosophical texts by Hans-Georg Gadamer, Edmund Husserl, Martin Heidegger, Maurice Merleau-Ponty and Don Ihde to examine the human–technology connection through the rhetorical and literary works in phenomenology to add depth to the interpretation. I also use work by Max van Manen to explore the teacher-scholar practice.

The phenomenological tradition is both a philosophy and a human science methodology embedded in discovering the essence of "the thing itself." The early focus on the birth of phenomenology stemmed from "an attempt to bring philosophy back from abstract metaphysical speculation wrapped up in pseudo-problems, in order to come into contact with the matters themselves, with concrete living experience" (Moran 2000, xiii). The things themselves, the matters themselves and the study of essences are all ways of pursuing phenomena for understanding an experience as it is lived. Phenomenology separates itself from other human science approaches because it existentially distinguishes between appearance, what we see, and the overall essence of an immediate experience (van Manen 1990).

The practice or foundation for human science research, based on written text, provides an interpretive model for the researcher. Gadamer explains that "For the interpreter to let himself be guided by the things themselves is obviously not a matter of a single 'conscientious' decision, but is the first, last and constant task" (2000, 267). This constant task occurs when the reading and the writing is translated, transcribed, illustrated and explained. He adds, "Beginning from the word as a means, we are asking what and how it communicates to the person who uses it" (412). To ask the right questions of the text is to bring the text into the open for interpretation. "A person skilled in the 'art' of questioning is the person who can prevent questions from being suppressed by the dominant opinion" (Gadamer 2000, 367). Gadamer's valuable advice is important because dominant opinion can suppress or cover up the technological texture.

The study of essences asks the question *what is . . . ?* to the world. As we relearn how to look at the world from different perspectives, we need a framework for the study of *What is*. Phenomenology as a methodological underpinning uses a reflective style of connecting with experience. The researcher engages with the content in a way that is similar to a reproduction, a casting of an image, or a certain aspect or impression of a phenomenon. The etymology of *reflect* is rooted in the Middle English *re*, meaning "to," and *f lectere*, meaning to "bend." The text creates a rebended, reproduced and realigned image to gain a certain impression. As I read over an anecdote, written reference or conversation transcription, I ask, what is its lived experience meaning or its interpretive point? The reflective writing process opens up ideas and brings meaning to the text. Many times a line or phrase will stand out within the text. These highlighted phrases help reveal understanding of the human–technology connection.

Phenomenological inquiry is often paired with the existential philosophy of language called hermeneutics, a theory and methodology that interpret written, verbal and oral communication. Hermeneutics brings forward themes for consideration and engagement. The process of analyzing the structural and thematic aspects of a particular experience brings out what phenomenologists call the essence of the experience to life, the qualities beyond mere facts and descriptions that make something what it is. Essences are not intrinsic features but ontological thresholds that we are motivated to grasp or understand (Ihde, 1999; Zhok, 2012). "The close relation between questioning and understanding is what gives the hermeneutic experience its true dimension" (374). Gadamer explains that the process is not to be understood as a methodology but as an "ontological structure of understanding" in the in-between-ness of the researcher's bond to the subject matter that comes from the text and the story that is being told between the interpreter and the text. Seeking the right words, the words that really linguistically name the thing being studied, is at the heart of this kind of research. Explains Gadamer,

> Through this interplay between words and the researcher, a practice of hermeneutic sensitivity occurs. The important thing is to be aware of one's own biases, so that the text can present itself in all its otherness and thus assert its own truth against one's own fore-meanings. (Gadamer 2000, 269)

Exploring hermeneutic phenomenology helps thread contextual understanding and provides a methodological underpinning for conversations and analysis of transcribed or written texts used in case study analysis. Unlike psychological analysis, for instance, a phenomenological analysis looks at a subject and his or her specific experience using digital technologies. Phenomenology brings forward the subjective human experience in deeply personal, historical, contextual, cross-cultural, political, and spiritual terms seeks the essence of a specific lived experience in a way that makes it real to us. Ihde clarifies,

> Phenomenology, in an initial and over-simple sense, may be characterized as a philosophical style that emphasizes a certain interpretation of human experience and that, in particular, concerns perception and bodily activity. Hermeneutics, on the other hand, arose out of the disciplines of textural interpretation and later a (Continental) type of language analysis. (1990, 23)

Studying and interpreting experiences constitutes the essence of something so that the structure of the lived experience is revealed to us in such a fashion that we are now able to grasp the nature and significance of the experience in new ways. The Canadian curriculum theorist Max van Manen, in his book *Researching Lived Experience: Human Science for an Action Sensitive Pedagogy* (1990), explores phenomenological writing and hermeneutic-phenomenology as a methodology for researchers. He says, "Phenomenological inquiry is not unlike an artistic endeavor, a creative attempt to somehow capture a certain phenomenon of life" (van Manen 1990, 39). The process is less concerned with the descriptive differences, or analysis of how many experiences share the same elements, and more concerned with the particular vantage point persons experience as they realize the essence of something in their day-to-day life.

At the root is the question of opening a phenomenon and unpacking it so it can be named. The essence of the *question* is to open up possibilities and keep them open for increased understanding. Part of the process is recognizing our own prejudices in light of other sources like interviews, stories, texts, songs and writings. This kind of methodology does not require us to dismiss our own experiences, but to relook at them in light of other experiences in their completeness.

Gadamer (2000) calls phenomenological research inquiry an acquisition of a horizon: "The concept of 'horizon' suggests itself because it expresses the superior breadth of vision that the person who is trying to understand must have. To acquire a horizon means that one learns to look beyond what is close at hand—not in order to look away from it but to see it better, within a larger whole and in truer proportions" (305). The research experience becomes an opportunity to examine a topic in the same way someone might look at a tapestry. The weaver puts the fabric up to the light and views it from all angles, noting the thickness or thinness, the heft, the texture, the lines and flaws not seen on the original technology called the weaver's loom. And it starts with horizon, which is situated in perspective. Metaphorical representations illustrate the perceptual backdrop of horizon in the lifeworld as a useful mode of perceptual analysis. Explains the French philosopher Maurice Merleau-Ponty, "The horizon is the point where my eyes fall on the landscape in front of me. The space-defining dimension can teach me what a point is only in virtue of the maintenance of a hither zone of corporeality from which to be seen, and round about it indeterminate horizons are the counterpart of this seeing" (2000, 102). The horizon, then, is the space defining the area where my lifeworld takes place. Van Manen explains, "Our lived experiences and the structures of meanings (themes) of our lives in terms of which these lived experiences can be described and interpreted constitute the immense complexity of the lifeworld" (1990, 101).

As we will see in the case studies later, the digital media experience occurs in our places, in and through (and on) our bodies, in time and through our relations and connections. A more situated naming of the experience comes through the existentials of the world; lived time, lived space, lived body, and lived relation. We experience digital media in these lived ways. For example, time might fly when we are gaming. Or, we might forget we are hungry when we are using social media until our body reminds us that we are hungry. We understand the lifeworld through these existential experiences and when technology is part of the experience, we live through it too.

The constant juxtapositioning between the source material gathered and mined for themes and the researcher's own experiences creates a kind of movement or engagement called *play*. Playing is something that children do involving games and fun activities. The idea of a "to-and-fro" motion or a movement without a specific end is central to the idea of play (Gadamer, 2000). This weaving, a kind of asking and answering of the gathered content, "absorbs the player into itself, and thus frees him from the burden of taking the initiative, which constitutes the actual strain of existence" (Gadamer, 105). When we experience this absorbing activity, we engage in the work so fully that we do not notice our corporeality. Playing also has been defined as a joyful, dancing experience. The Middle Dutch word *playen* suggest practice, cultivation, and attention (Barnhart, 2001). Play is a foundational concept for the phenomenological researcher because the questioning lays bare prejudices or parts of the phenomenon we cannot or will not personally see. This is what some psychologists call *flow*, the happy condition of encountering optimal experience. Concentration is so intense that there is no attention left over to think about anything irrelevant, or to worry about problems. Self-consciousness disappears, and the sense of time becomes distorted. (Csikszentmihalyi 1990, 71) Exploring personal written accounts, close observation, biographical works, transcribed audio or visual conversations and other sources are ways to gather data for the researcher but also a way to play with the materials to cultivate and attend to specifics.

Phenomenology can also be described as "the study of 'persons,' or beings that have 'consciousness' and that 'act purposefully' in and on the world by creating objects of 'meaning' that are 'expressions' of how human beings exist in the world" (van Manen 1990, 4), "and this type of methodology is often interchanged with the movements of phenomenology and hermeneutics" (2). The emphasis on interpretation within the phenomenological tradition allows lifeworld experiences to be uncovered and explored to make central the experience of the body and perception. The focus of hermeneutics helps to name these lived experiences raised in a phenomenological study. "In other words, phenomenology does not produce empirical or theoretical observations or accounts. Instead, it offers accounts of experienced space, time, body and human relation as we live them" (van Manen 1990, 184). In almost the same way as a documentary reveals a phenomenon through a visual study, the researcher begins to gather text that explores and questions the phenomenon.

Now that a brief grounding of phenomenology has been explored, it is important to further flesh-out postphenomenology as the main framework for the digital media cases highlighting the screen, digital sound, music, photo manipulation, datamining, aggregate news and self-tracking that come in the second section, "Feeling The Weave." Phenomenology is inherently embedded in postphenomenology but does not provide the interrelated ontology we need to thoroughly interrogate digital media. Postphenomenology is what happens after the phenomenon has been discovered and explored in a variety of phenomenological ways. The first task was to begin laying a deeper groundwork for understanding phenomenology, so the postphenomenological turn makes sense.

POSTPHENOMENOLOGY

In the first decade of the twenty-first century, scholars started actively exploring philosophy of science and technology, also known as Technoscience Studies, to search for ways to differently understand the instrument-human-world relationship, or being-in-the-world-with-technology. The aim was to explore the technological texture of the human–technology connection. A group of scholars turned to studying ways to situate and assess technological experience from the user perspective. The initial focus was largely on medical imaging and diagnostic technologies and design experiments for both the arts and sciences. The postphenomenological researcher Robert Rosenberger notes, "Postphenomenologists attempt to combine the philosophical traditions phenomenology and pragmatism; they investigate issues of human relations to technology; and they place an emphasis upon the analysis of concrete case studies" (2008, 64). This applied philosophy has contributed to a wide range of research in fields across the humanities and social sciences (Ihde, 2008; Riis, 2010). The aim of this study is to bring media and related studies into the conversation to interpret the whole of digital media differently. As described by Don Ihde, who is often called the father of postphenomenology,

> With the now growing body of postphenomenological research projects, one can also see that postphenomenology is not open to the critique of classical

phenomenology as being anti-scientific. Rather, in a significant number of cases, postphenomenological researchers have actually become engaged in facets of experiment design and engagement with the relevant scientific communities. They investigate phenomena in what I call the "R&D" or research and development sites where initial considerations and critiques can be helpful. (2008, 7)

Ulrich Beck, in his essay "Politics of Risk Society," pushes the importance of engaging in critiques of the current climate in R&D. "In the age of risk, society becomes a laboratory with nobody responsible for the outcomes of experiments. The private sphere's creation of risks means that it can no longer be considered apolitical" (1998, 588). Human relations to technology are central to contemporary R&D. Investment in research, product and design innovation and development; manufacturing and factory management and research priorities all become attached to both politics and culture. This man-made hybrid world, where nature and culture converge, combines "politics, ethics' mathematics, mass media, technologies, cultural definitions and precepts. In risk society modern society becomes reflexive, that is, becomes both an issue and a problem for itself" (588). A postphenomenological framework can be central in plotting trajectories and identifying multistable variations to make sense and bring direction to these issues.

Postphenomenology owes its roots to phenomenology but reflects historical changes in the twenty-first century turn toward case study research as a way of informing a whole host of technological studies. Postphenomenology is described as a *non-subjectivistic and interrelational phenomenology* that is useful to study the human–technology experience. Ihde explains that postphenomenology takes a "step away from generalizations about *technology uberhaupt* and a step into examination of *technologies and their peculiarities* . . . an appreciation of the multidimensionalities of technologies as *material cultures* within a *lifeworld*. And this move is a step into the style of much science studies, which deals with case studies" (1990, 22).

Postphenomenology is often put into a box with the variety of other postphilosophies, like post modernism and post-structuralism. While there might be similarities among them, I'd like to separate postphenomenology from the philosophical pack specifically because of the relational sensitivity, interest in peculiarities and a focus on multidimensionalities of technologies in material culture. The human in the human–technology connection is ontologically related to the lifeworld but both transform to become co-shaped within the connection. Digital media and its ubiquitous devices are embedded in this environment and are part of the human lifeworld experience. But how to study it? Explains Ihde, "Postphenomenology begins with what is familiar, but then begins to move beyond that into more radical variational possibility" (2006, 289). This framework is an appropriate and fruitful place to explore digital media. Specifically, postphenomenology's variational theory, multistability, macroperception and microperception are key concepts that probe technology in a different sort of way.

Variational Theory

The first step of postphenomenology is identifying variations in the reading/ use of the technology. Variational analyses can show multistabilities, those distinctions and differences that may or may not be obvious at first glance, but also can be an analysis framework in itself. Variations "yield both deeper and more rigorous analyses of such illusions than mere empirical or psychological methods" (Ihde 2006, 12). Rosenberger notes, "According to Ihde, phenomenological analysis should proceed with the creative exploration of different interpretations, or variations, as he calls them" (2011, 9). With this sort of analysis, the range and limits of the possibilities for interpretation can be investigated. In the case of an image of the real world (such as an image in science), the viewer does not understand each different variant to correctly explain the real world phenomena captured within the image; the alternatives are perceptions of the meditating technology, the image. Ihde explores the following list as possibilities for identifying variational interpretations.

- Materiality of the technologies
- Bodily techniques of use
- Cultural context of the practice
- Embodiment in trained practice
- The "appearance of differently structured lifeworlds relative to historical cultures and environments" (2009, 19)

The Swedish researcher Anette Forss notes, "Human–technology–world relations cannot, however, be reduced to one thing and instead must be untangled through variational analysis to elicit and describe perceptive multistability" (2012, 298). The nod to technological mediation provides a foundation into the digital media world of entangled embodied, mediated experience that is not only reduced to designed functions. For digital media, exploring perceptions, interpretations, use, context, content, materiality, cultures, environments, structures, workflows, practices and use can yield multiple options and possibilities that open up understanding in a variety of ways for exploring technological texture. Variations can serve the everyday needs of the user to entertain, to educate and to inform, but also to shape society, shape communication and shape political discourse and activity.

Multistability

Multistability is one of the central overarching concepts of postphenomenology. This foundational theoretical underpinning shares kinship with the human learning concept of gestalt theory, a way to group a problem perceptually based on proximity, similarity, closure and simplicity. Postphenomenology's multi-perspectival, relational analyses opens the weave in the human–technology connection to see the technology differently and from a variety of interpretations and entanglements. Multistability explains that the very structure of technologies are multistable with respect to uses, cultural embeddedness, politics and ethics, among other things. But there is more to it. Imaginative multistabilities and practical stabilities both build case studies that analyze variability. "First, multistability is an empirically testable hypothesis about how several stable patterns of the same object can be perceived from the first person perspective." Practical multistability is based on the hypothesis that human bodies and technologies are entangled in the lifeworld and includes sets of concepts and criteria that can be used to describe some of these entanglements. The descriptions are used to shed light on the role of these entanglements in framing our aesthetic, moral and political values and the possible ways of improving the benefits and sustainability of technology design (Whyte, 2015).

Kyle Whyte's distinction between the concept of imaginative and practical multistabilities starts with the idea of stability, "an umbrella term that can refer to anything perceived as having a constant pattern, from the constancy of images, to practices, to technologies and so on. Anything that is stable comes across to us as having at least one of the following: a particular look, a particular way of acting or a particular use" (70). This book includes both imaginative and practical multistable analyses and could be considered one large digital media case of multistabilities. But, broken down, each digital media has its own embodiment, cultural embeddedness, variations and different human–technology connection. These non-neutral contexts and practices are the focus of this book. As a researcher studying a subject through postphenomenology, I might find that I am stuck on only one or two different, stable possibilities. I see a specific subject only these ways and I move between the two but am unable to see a third variation. The idea is often illustrated by the Necker Cube experiment (Ihde 1986, 102–08). We see either the back or the left of the cube. We wonder, which are the stable presentations and what new ones are revealed as we continue to study them? What is actual and what is potential?

Postphenomenology is "focused on familiar uses and cultural norms. Within multistability there also lies *trajectories, and* not just any trajectory, but partial trajectories" (Ihde, 2002). Alternative views and trajectory paths strengthen multistable thinking and expand ideas. Human–technology connection within a pluralizing of cultures and possibilities brings out multistable possibilities. Rosenberger explains it this way: "Though we may typically embody a technology in a specific way, there are always other stable and coherent ways that the relation can occur. For instance, a typical glass bottle can be used for purposes other than holding liquid. It can be used as a vase for a short-stemmed flower. It can be used as a base for launching 'bottle rocket' fireworks" (2009, 175–76). Digital technologies are also used differently. A smart phone provides digital communication between people. But this digital-media device inherently carries many uses and trajectories specifically designed for multiple uses. Users want one device that does it all. The open space on the screen indicates that there is room for more apps. The space for more becomes an indication that the industry plans to grow to fill that open space. Additionally, each phone is it's own unique tool. While each technology has the same beginnings, the built in and inherent multiplicity is potentially endless. One person's phone is a journalist's device but others may see that same technology as a gamer's playground or a social media maven's work surface. Different apps multiply and personalize use and proliferate the multiplicity. No phone is the same. Smart phones provide gaming entertainment, music listening, geographic tracking, self-tracking, photo manipulation, media viewing and a variety of other app-related trajectories. The device can also carry relational roles of companion, parent and observer (Rosenberger, 2012; Spicer, 2014; Wellner, 2014 and 2015). And a smart phone also becomes a paperweight once the battery has run dead. "Multistability is neither a variant of relativism nor of social constructivism. It, instead, unifies two philosophical schools of perception seen as interrelated; the sensory bodily perceptualism of Husserl and Merleau-Ponty (called microperception) and the cultural perception found in Heidegger and Foucault (called macroperception)" (Forss 2012).

How do we get digital media to speak, to reveal the possibilities? While classical phenomenology would work to reveal essences, a more pragmatic or postphenomenological approach allows technologies to evidence themselves through the human-world connection. In postphenomenology, this is called the *empirical turn*, defined as investigating how the things themselves mediate being-in-the-world. Conversations with those who use digital media show various trajectories and user connections. Written texts show hermeneutic-phenomenological renderings. Other variations produce multistabilities, not essential structures or essences (Ihde, 2009). Whyte explains another concept that is central to postphenomenological variations called pivoting. This explanation is essential to understanding multistability. In his essay, "What is Multistability: A Theory of the Keystone Concept of Postphenomnological Research," Whyte concisely explains "To pivot on the artifact requires that we assume the identity of the artifact remains constant across the variations. This pivot stresses the degree to which the material of the artifact and human intentions can create different uses. To pivot on the practice requires that the identity of the practice remains stable. This pivot stresses the importance of social, cultural and historical context in shaping what people do with technologies. Each pivot requires that we assume that something remains constant and that the reader will not question the identity" (Whyte 2015, 76). He adds, "Standards bring into being the stabilities of the human–technology entanglement. Standards also are mediated by the social and cultural context of those who create them" (Whyte's interpretation of Verbeek 2015, 77). The idea of pivots explain the constants against the variants for the case studies in the second section of the book but pivot points will not specifically be identified in each case.

Macro- and Micro-Perception

Multistability shows itself through micro- and macro-perceptions. *Microperception* is sensory, fundamental and reflexive to bodily position. *Macroperception* is what contexts micro-perception and as contexts can vary it yields cultural diversity and thus gives way to an understanding of perception as polymorphic. "Through this analytical distinction the expanded phenomenology of perception which links micro- and macro-dimensions can give clues to the shape" of multistability (Hasse 2008, 46). Each microperception happens within its macroperceptual environment. The body-sensory (micro) and the perceptual (field) create ambiguity. Both are important for human–technology connection with digital media. Concludes Ihde, "[M]acroperception is what contexts the microperceptual" (76) through a hermeneutic or historico-cultural reframing. These two "ways of seeing" produce variations through multistability that juxtapose to reveal different things about our technological texturing through digital media.

Another important distinction for digital media focuses on the body. One way to think of embodiment and its link to digital media is through Ihde's (2002) Body One and Body Two explanation. Body One refers to our everyday embodied state. This is our lived body. Body Two serves as our social and cultural body. This body is pushed by cultural norms about how we wear our technology and when "bring your own device (BYOD)" is initiated at work or school. Navigating across this terrain and crossing in and through it is a third space or area or in-between. This is the human–technology connection. Another way to think about corporeal experience is through the idea of progressive variations of *affect*. Critical-cultural scholar Greg Seigworth uses the work of Gilles Deleuze and Felix Guattari to explore *affect*, the emotions and subjective capacity of our bodies to experience the lifeworld from *affection* to *affectus* to the immanently expressive world that expands or widens to all of the moments that humans go through, to explore the wholeness of experience. Affect Theory seems to get at the expressions of the world made through the human–technology connection with digital media. The idea of "world-as-expression" (2005, 168) through digital media can push media effects to media affects and provide a richness to the study of macro and micro-perceptions. Seigworth notes, "Locating the plane of immanence is not unlike discovering the intricate weave and meshings of the whole fabric of cloth, constantly moving, folding and curling upon itself even as it stretches beyond and below the horizon of the social field (without ever separating from it or departing it" (2005, 168). Which body and what experience does digital media announce or is that third space, that plane of immanence, always at play?

A TRANSPARENT TEXTURE

My technology extends my body into the world in different ways through digital media. To alter perception while using these devices, the "I," acting as a user, might think of the instrument as enlisted in the service of my actions, not as a model of my thought (Rothenberg, 1993). There are certain "taken-for-granted" assumptions when using technology. We might not consciously think through these ideas, but somewhere in the human–technology connection, we consider the *life* of the technology. Notions of personification or anthropomorphism change our connections and relationships to our technologies. We might not believe digital media technology is alive, but devices are called *smart* and users note that technology "hates them." Researchers Bartneck, Hoek, Mubin, and Mahmud (2007) studied connections between an iCat robot and participants in their study, by creating a situation where the robot and the study participant played a game together. At the end of the game the participants were asked to turn the robot off. While the participant tried to turn off the robot, the iCat begged, in either an agreeable or not agreeable tone, not to be turned off. The researchers concluded that the study

> suggests that intelligent robots are perceived to be more alive. The Manus Manum Lavet rule can only apply if the switching off is perceived as having negative consequences for the robot. Switching off a robot can only be considered a negative event if the robot is to some degree alive. If a robot would not be perceived as being alive then switching it off would not matter (221).

Embodied technology like the iCat introduces many ethical questions. Virtual and augmented reality also blur the line between humans and technologies. The tether to our digital media is very real. The need to grant human characteristics to inhuman objects reigns, and I again see the complexity of the technology I am entwined in. Minimal human-like behaviors are needed to trigger a social response.

> Yet our relationship to the anthropomorphized machine is complex: When asked directly whether a computer can think, many would say 'no,' although in actuality they interact with the machine as if it were a thinking being, attributing violation to it and reacting to its 'opinions' much as they might another person's. (Donath 2000, 301)

The sense, then, of human–technology connection brings the user into an intertwining of many feelings, including disappointment and elation when the appropriate connection is achieved or lost. Humans make sense of the world by creating different kinds of social categories in their relation with other humans. The need for this categorical kind of relating seeps into our lived experience with our digital media. The human-like classification helps make meaning of the human–technology connection, which adds a new layer as we think about the crux of this study, not to overlook the familiar and common to the point where they seem neutral and absent in the lifeworld.

The first philosophical section of this chapter provided grounding and general knowledge about the Continental European tradition called phenomenology. This underpinning opens the way for postphenomenology by exploring the contributions that the lived experience research has to postphenomenology, namely through the range of possibilities for seeing variations and multistabilities in digital media's human–technology connection. Postphenomenology is the main framework of analysis through the second section of the book, "Feeling the Weave," where the technological texture is illustrated through case studies. When it comes to studying digital media, a less essentialist view can bring forward more use contexts to consider for opening up the reified and defined set of attributes to name multiple identities and functions. Variational theory teases out important alternatives to thinking, to revealing the prevalent and taken-for-granted technological texture of digital media with regards to historical and socio-cultural considerations. The notion of multistability will prove useful for drawing out how things themselves mediate being-in-the-world. Multistability explores the often-evident use contexts or valid interpretations of how digital media might be used, and then pushes other additional and valid considerations. The variations, macro and micro perceptions, and multistabilities are different for each case study, but each works to revealing patterns and alternative understandings. Concepts like being-in-the-world, emobodiment, and perception clarify the digital experience.

We now move to an exploration of one of the foundational technological compositions of digital media, the screen. This case digs deeper, as do the others, to discover and illustrate aspects and variations of our technologically textured world to further explore the human–technology connection. The topics of digital sound and music, photo manipulation, data mining, aggregate news, and self-tracking establish the varied and non-neutral digital media patterns of our technological texture. As a whole, they provide a starting point and illustrate variations that challenge standard stabilities of the human–technology connection and push the perception that digital media does not lack effect.

NOTE

1. See also Ihde's *Postphenomenology: Essays in the Postmodern Context* (1995), *Postphenomenology and Technoscience: The Peking University Lectures* (2009), *Heidegger's Technologies: Postphenomenological Perspectives* (2010), *Experimental Phenomenology, Second Edition: Multistabilities* (2012), Evan Selinger's *Postphenomenology: A Critical Companion to Ihde* (2006) and Jan Kyrre, Berg O. Friis and Robert P. Crease's *Technoscience and Postphenomenology: The Manhattan Papers* (2015).

II

Feeling the Weave

Chapter Four

Case: The Screen

Our technologically textured world is worth studying to reveal different ways of recognizing impact through the interrelated ontology of human–technology connection. This case study brings the focus to several metaphors that identify within them, variations of one of the most foundational parts of digital media, the screen. Variations in sociological and cultural and political movements (macro-perception) and sensory and embodied experiences (micro-perception) are also considered as two "ways of seeing" variation in the lifeworld. Many times research on the screen migrates toward talk of images, but this case investigates the screen as a foundational technological element of digital media, and explores the relational ontology of the screen as technological reciprocity for human–technology connection.[1]

A great way to start reflecting on this case study is through an understanding of the word *screen*, which has many variational, language use patterns and meanings. A screen closes off, partitions and sections off an area. This word can mean an apparatus that protects from heat or fire, an open mesh for sifting, mesh on a window frame to protect against insects or a cover, shield or a surface where images are shown. Examples like protecting and shielding suggest closing off, and sifting suggests partial removal, protection or examination. A screen is also a surface for projecting or receiving information. How might those etymological roots tell the tale of the screen's prominent status and significance for digital media? We will explore the representation of the word screen and metaphorical explanations of screen type use, and then discuss the human–technology connection through postphenomenology.

This case study focuses on the postphenomenological concept of variational theory discussed in chapter 3. The idea is to illustrate distinctions and differences that might not initially be obvious. These observations uncover micro- and macro-perceptions that co-shape socially and culturally embedded understanding. Together, variations and perceptions work to produce a rendering that begins to name the screen experience. All microperceptions are linked with macroperceptions, so the micro screen experience is explored first and then the macroperception is considered. But first, let's look at instrumentally mediated perception and metaphor as a backdrop for a post-phenomenological investigation of the screen.

CONTEXTUAL BACKDROP

In the early days of television, screens were part of a larger instrument called display technology. Converting information to fired electrons and onto a luminescent phosphor-covered screen created early TV displays. The result was a visual image. Engineers created waveforms and other informational displays using this technology, and numbers, images and graphic representations soon followed. Cathode ray tubes replaced phosphor-covered screens to become the standard for mass-produced televisions, medical equipment and computers. In recent years the LCD, plasma and 4K ultra screens have become popular display technologies. In the history of communication technology, screens have gone from small, to very large and back to small again. On any given day in contemporary society, we can see all kinds of screens on the latest popular technology. Digital media relies heavily on digital images, which require display, present a perspective and are often associated with a screen.

With digital screening, there is no longer a fixed combining between image and image carrier (as in the photographic print) but rather a temporary alliance of the image with the place of its apparition, the screen. The image is no longer tied to a specific medium of production and is open to a variety of display formats. The impression is multi-platform and comes to be defined through that platform or display rather than through an inherent characteristic or ontology. In short, digital media content could be initially viewed on a mobile device and then finished on a tablet, laptop or larger home screen. We may not call it watching TV anymore because we watch a specific kind of content (the name of the program) or a disseminating brand of curated subject matter (Netflix, Xfinity, Hulu, YouTube). It is still highly individual in nature, with earbuds or headphones connected when in public, and door closed to watch in private.

Screens are anecdotally considered and discussed in mainstream media, but the conversation may be relegated to arguments about too much or not enough screen time, or how large or small a screen should be. It seems important to acknowledge that the digital media industry is always working to create environments where the instrument is less perceived and increasingly replaced by a sensor that acts as a conduit or conductor of information strapped to the human body or surrounding lifeworld. In fact, this is the specific goal of much of virtual reality and augmented reality's research and development focus. "Mixed reality applications are already starting to engage users and players to interact with the city environment. The combination of virtual content with citywide sensors presents a perfect combination for the creation of an interactive Augmented City, where citizens will experience more intense and immersive experiences" (Nóbrega et al, 2014). While these technologies continue to expand and improve experience through ideas like Augmented City and wearables, the screen endures.

I view the screen in a forward stance. The Latin root *stans*, a resting place, and *stare*, to stand (Barnhart, 2001), suggests a position for viewing the horizon. Many technology users say they "sit in front of" the screen. The Italian root *stanza* suggests a stopping place as a stance. Is the front of the digital media the location where the body stops while the mind is engaged within the computer? Stance suggests an intellectual point of view or posture toward something. What am I saying when I stop in front of the screen? For me, to be in front of the computer is to face this kind of spatial arrangement and orientation. I make everything look the same way each time. It is a ritual born of habit from the very first day I stared at the computer for verification that we were connecting. This habit of orientation within the software through the screen is a constant negotiation of lived space. This is lived space, sacred space and personal space, fit to the body like sitting into a comfortable chair. "Our relationship unfolds in the space created by our technologically supplemented bodies" (Leder 1990, 34) and extends our personal and natural bodies. Without that screen, I could not properly translate the software and pen to crop and paste my visual story together. Without looking at the screen, I could not view the digital world.

When I look through my device screen, am I losing the horizon or gaining the horizon? I gain access to new worlds, but my body seems absent/absorbed from formal engagement with the lived world. What is my cue to look for the screen on every new device I encounter? Do I just know to look, because the semiotic cues are embedded in my humanity, my user mentality or my tool being? Can I live in multiple enfolding horizons, like the multiple windows and tabs I set on my desktop. Can my technology world be one horizon and my lifeworld be another horizon? Or have they collapsed together? The engagement with screened technology is rooted in horizon and horizon is rooted in perception. Indeed, as Dave Abrams says, "from the perception of my bodily senses, there is no thing that appears as a completely determinate or finished object. Each thing, each entity that my body sees, presents some face or facet of itself to my gaze while withholding other aspects from view" (1996, 50). This is the perceptual experience of using screens.

Chapter 4

PERCEPTUAL MEDIATION

One of the seminal philosophical works specifically aimed at understanding visual technology phenomenologically is Vivian Sobchack's *The Address of the Eye: A Phenomenology of Film Experience* (1992). Sobchack's philosophical account of the lived experience of film viewing explores the important concept called "instrument-mediated perception" (172), as the in-between for the human–technology connection. Sobchack explains the filmmaker experience as duel living within the "lens-world junction" (175) and the "monitor-world junction" (175). When a film is seen through a projector, it can be viewed in front of the projector or behind the projector. In both instances, the film can be seen, but the technology is absorbed into the experience if the projector is behind the viewer and noticed if it is in front of the viewer. Instrument-mediated perception allows a change in thinking regarding the human–technology connection. When might the digital media screen be absorbed and when is it noticed? The technology can be in the room with the viewer but play very different roles in the mediated experience based on location within the viewing experience. For instance, a user wearing virtual-reality glasses experiences the game differently from a user playing with a remote in hand and a display monitor on a wall. A driver with a heads-up display does not look down from the road to check speed or temperature. The continuum between highly apparent and absolutely transparency rests on perception and the body as a key locator in the perception.

In their 2004 study, "The Ontological Screening of Contemporary Life: A Phenomenological Analysis of Screens," Lucas D. Introna and Fernando M. Ilharco study the phenomenological idea of what it means when we engage with a screened surface. They explain that a screen shows an already prepackaged version of content that has been specifically created to be shown in a presentation. Their study describes that screens are looked at not because they are screens, but because we wish to see what is on them. They explain that our lives are ordered to seek what is presented on the screen, and the design and orientation of the screen are already part of our everydayness. This use pattern is not neutral. It changes our behavior by its existence. Introna and Ilharco explain, "The screen, in order to be a screen, assumes in its screening an already shared referential whole of language, symbols, practices, beliefs, values and so forth, for its ongoing being . . . it implies a sharing and co-constitution of a form of life as the assumed possibility for the screen to be a 'screen'" (238). This co-constitution or co-shaping is the connection between human and technology. And if we are increasingly and referentially attuned to screens, the use becomes a kind of learned neutrality. We are co-constituted to the point where we do not even know that we are looking at a screen. Exploring metaphorical variations provides an analytical framework for better understanding human–technology connection.

METAPHORICAL INTERLACE

Anne Friedberg explores the metaphorical nuances of the screen in *The Virtual Window: From Alberti to Microsoft* (2006). Her ideas extend Sobchack's work into more recent digital technologies. She firmly illustrates the use of metaphor as the one tool that is developing language that illustrates our technology experience. Lakoff and Johnson (1999) describe how metaphors explain, among other notions, space and movement. The figurative notions of a machine/computer metaphor highlight the idea that "the mind is the software and the brain is the hardware" (252). Metaphors "attend to some likeness . . . between two or more things" (Davidson 1981, 202) and allows an "experience [of] one type of thing in terms of another" (Lakoff and Johnson 1980, 5). Embodiment of primary metaphors in software design and screen use helps bridge the gap to better understand the technological environment. Interface creators are moving away from iconic graphics folders for storing files, floppy discs for save buttons and trashcans for throwing files away—that overtly visually describe metaphorical use. App icons are becoming more simplified and streamlined in design. The newly embraced flat design sheds cartoony 3D graphics in favor of a more simplified approach without textures, shadows, or gradient style. This design also looks good on very small screens, where nuanced shadows and detail are rarely noticed.

Early attempts to understand this screened space made philosophers specifically think about what "place" users were in as they reached out to the online "cyberspace" environment. Naming kinds of cyberspace, like two-dimensional text based Barlovian cyberspace, created language for better description. Digital media technologies alter symbols that are signified by a virtual icon or program and they may be represented on device screens by icons that illustrate shapes and locations (Brey 1998). Metaphorical illustration through graphics situates the place beyond the screen and explains representation and orientation metaphors for the human–technology connection. Metaphors are imbedded in visual design and will be part of the newer interfaces moving forward. Metaphors also move into macroperceptual understanding because they say much about social and cultural meaning-making experience in the lifeworld.

An "orientational metaphor" (Lakoff and Johnson 1980, 14) explores situated embodiment knowledge of familiarity and habit, allowing the participant to reveal his or her relation to technology through culture and social habits. At one point in time, a screen worn on the body or held in the hands would have seemed an oddity on public transit. Watching visual entertainment on a cell phone would have been unheard of. And listening to people seemingly talk to themselves through on-the-ear headsets would have turned heads. Noticing someone using Google Glass might cause others to stop and stare. Different countries have integrated technologies at different paces and some cultures skip certain embodiment stages all together depending on when the technology is adopted. The macroperception of a screen's use is largely defined by the technology used in that culture. All newly invented technologies tend to be viewed as oddities and fads and are either integrated into the culture or rejected for a variety of social, cultural and political reasons.

The screen is oriented to provide a look inside a device. To truly be inside is to know a place intimately. Knowing is not just knowing, but personally knowing, as opposed to tarrying alongside. Reoccurrent experience leads to the formation of categories, which are experiential gestalts that promote meaningful perceptions with those natural dimensions. Such gestalts define coherence in our experience. We understand our experience directly when we see it as being structured coherently in terms of gestalts that have emerged directly from interaction in our environment (Lakoff and Johnson 1980). This important idea suggests that to use something is to know it. "[A]s the technologies and artifacts become more complicated and less transparent e.g. computers), their role in affecting values and cultures becomes greater," (Ihde and Selinger 2003, 184). Just as we can alter our sense of our personal temperature by viewing a thermometer, the translatory tool that visualizes temperature, the artist can visualize the work within the digital environment. "Each of us is a container, with a bounded surface and an in-out orientation. We project our own in-out orientation onto other physical objects that are bounded by surfaces . . . There are few human instincts more basic than territoriality . . . We conceptualize our visual field as a container and conceptualize what we see as being inside it. Even the term, 'visual field' suggests this" (Lakoff and Johnson 1980, 28–30). Lived orientation is bounded by surfaces and humanness looks for the faces in the crowd. The screen, that possible conceptualization of a technological face, negotiates and translates connectedness with the instrumentally mediated and tooled world. It is from quantifying and partitioning through the use of metaphor, that humans make sense of their world, because they are bounded by their surface (1980). Humans are always in relation to and with technology in the lived world, and this alters human praxis.

Frame Metaphor

Friedberg calls that framed boundary an ontological cut because of the complete ending of the boundaries. She notes, "The frame itself carries with it some subjective consequences. Like perspective, both the window and the frame serve as philosophical paradigms and aesthetic devices" (2006, 11). Digital media always has some framed structure for viewing something through, or to surround to bring into focus or enhance. An early use of *frame* is rooted in the Old English idea of profiting, making helpful or joining together, as in framing a house. Old English (Barnhart 2001, 800) suggests that *frame* meant to promote or to benefit, (1000) move forward, make progress, promote or influence. Old Saxon use of frame notes performance, composition, design or fashion, and the Old Icelandic word use means advancement (1250) or a plan.

The frame is an essential component to early visual images. Artists frame paintings, and filmmakers capture a frame of film. How might the digital media frame promote or exclude the image? Friedberg notes, "The moving image (of frames) produces a complex and fractured representation of space and time. And once two or more moving images are included within a single frame, split screen or multiple screen films, inset screens on televisions, multiple windows on the computer screen, an even more fractured spatiotemporal representational system emerges" (93). Friedberg explains that our everyday experiences with frames, the actual material frames of our technology, illustrate the frames dominance in our society. Frames have edges. Frames can enfold themselves within frames. "The frame of the screen is a closed system, a primary container for inset secondary and tertiary frames that may recede in *mise en abyme* but also converge to reunite within a grander but still bounded frame" (241). Aspect ratios and size may change, but as Heidegger explores in his essay *The Age of the World Picture*, the implied frame in the picture is a fixing of position (1977). His notion of frame, *Das Gestell*, as a metaphor for representational thought, organizes perception, sets everything in place, and orders the world (95). The idea of an embodied fixed position opens the notion of enframing, which turns the world into objects, into a standing reserve, awaiting its representation and introduces the concept of using tools and technology exclusively and always in-order-to do something. The premeditated use has implications for human–technology connection. Or as Friedberg notes, the world, ready for its close up.

Window Metaphor

But the screen also brings us into a different kind of frame—a window. Philosophy has always had an interest in thinking about windows, portals and horizons. If to look at an object is to inhabit it, then to look at a computer screen is also a way to inhabit, gaze into, or otherwise grasp the presentation. The metaphor of the computer window is synonymous with the windows platform. Although windows can be open, metaphorically, they cannot reveal their innerworkings. In fact, almost all interfaces are designed to hide the code. Windows allow for screen-based multitasking (Friedberg 2006, 233). The participant can use multiple screens, engage in multiple activities, and mix business (computation, checking work email, downloading documents, rendering video or illustrator work), with pleasure (social media, games and chat).

Now that the screen has been explored through language and metaphor, we can move toward a postphenomenological analysis of one of the foundational elements of digital media. When I first look at a screen, I look right through it and practically ignore it. My horizon is beyond its materiality. I am looking to what it can offer, like information or a story. I have to really pay attention to see the screen as a material object, to see my reflection on it and then the imaged digital content on/within it. When I look to the screen I am engulfed in the experience; my perception shifts and any foregrounded technology is backgrounded. If the content is compelling, dramatic or otherwise engrossing, I become unaware of my body and become absorbed within the technology. This is especially clear when I watch TV or play a game. Double perception is at play. The screen is a paramount part of contemporary visual technology, but the human gaze goes right through the screen that shares the visual content. That is the way it is designed. Shiny surfaces reflect and distract, so designers have worked hard to engineer matte finishes that remove distraction and increase comfort for reading, viewing and experiencing.

CONNECTING HUMAN AND TECHNOLOGY

Sometimes, a screen can be an invitation for spectator viewing. "Spectators are," writes Gadamer, "set at an absolute aesthetic distance in a true sense, for it [the screen] signifies distance, the distance necessary for seeing, and thus makes possible a genuine and comprehensive participation in what is present before" (2000, 127). But as the screen moves closer, the perception changes. As spectators, now increasingly participants, the perceptual seeing might not be completely absorbing because there is always an "eco-focus" that lets the lived body know that this perception is not wholly his or her own and that there is something else there, standing in the in-between (Sobchack 1992, 178). "Although I can see through or according to or because of the screen, I cannot see like or as a machine; I cannot see except against the ground of my human lived-body and I cannot see unintentionally" (183). The eco-focus might be less distinct, but there is still the screen type material in between. Instrument-mediated perception, a genuine lived perception, still only has partial transparency.

Digital media content is understood through the screen. The "on" button of a piece of technology is pushed, the screen flickers or boots and the user stares expectantly at the screen before making the next move. The work cannot progress if the screen stays blank. The screen of image technology bridges the gap between what can be humanly seen and what might be there that cannot be seen. A screen is a non-neutral artifact because it plays a role in the formation and change of our human values in the world (Ihde and

Selinger, 2003). The screen is the "face" of the technological interface. It is something to look at and relate to.

> We are drawn into faces and respond to them, because they are turned to the world and themselves. The face indicates the regard of the being, directed toward things, creatures and events in the world around it. (Mazis 2008, 99)

Pushing buttons, a text pad or keyboard, invites the "look" to the screened face for recognition when we use our devices.

Most digital devices today include a screen of some sort as part of its interface, to "translate" the technology code. We type and swipe, text and browse, and view our way through our days peering toward some small, medium, large or extra large flat surface connected in some way to a digital device. Thinking through this presence in a very concrete existential way, face-to-screen, can be useful to further understand the human–technology connection. When technology becomes part of the general intentionality relations for perception, the metaphorical model shifts to the instrumental "Human-technology-world" relation highlighted in chapter 1 (Ihde 1990, 85). Technology shifts to a co-constitutionally mediating position. The relational connection is no longer face-to-face in the world, as in a human relation, but face-to-screen. The screen is the face of the technological interface. It is something familiar to look at and relate to, an opening toward the interface, the graphical connection that makes meaning between face and screen. Notes Ihde, "In this interconnection of embodied being and environing world, what happens in the interface is what is important. At least that is the way a phenomenological perspective takes shape" (2002, 87). Even when the connection is mediated through bodily contact with technology, like Google Glass, a smart watch or a remote "shake," a specific place is designed to focus the embodied input. In this context the screen becomes the embodied center for the user, with the interface as mediator and screen as translator. A positive or pleasing result reflects a happy face. Pushing it further, the screen can be thought of as a "subset of the face, as a 'quasi-face'" (Wellner 2014, 2). The face, or perceived screen as face, plays an important role in the relation with the Technological Other, that digital other that I know and connect with (Irwin 2005). Galit Wellner, in her essay "The Quasi-Face of the Cell phone: Rethinking Alterity and Screens," explains that screens can allow expansive and conversational or reduced versions of the face. It seems like technology gives just enough face to promote relationships. She explains that the screen "acts like a face that requires a response, but it is not a face. It is a quasi-face" (2014, 13). This referential relation extends and makes sense to the user.

If Merleau-Ponty's sense of plunging into the things perceived is connected with the quasi-faceness, then the screen may become either a poorly lit entry way, portal, opening or frame to plunge into in an act of absorption. Wellner adds, "like the face serves as an inter-face point between humans, the screen serves as an interface point between humans and technologies" (2000, 15). This further illustrates the screen's importance for human–technology connection. Glen Mazis, in his book *Earthbodies: Rediscovering Our Planetary Senses*, explores what he calls the "screening off" experience, or removal of the body-to-body fleshly communicative possibilities of technology. He says that "part of the sensual 'thinness' of projected photographic images, animated scenes, or even projected video clips is a screening of the material medium needed to carry the emotional depth . . ." (2002, 164). We are once, possibly twice, removed, depending on how many times the content has been screened through the production process before it reaches the spectator.

Digital media allows us to view content on big and small screens and on multiple screens at the same time as we multitask between our laptop, entertainment screen, and smart phone as separate pieces of equipment. Or we can view our TV programming, or messages and our documents on different browsers and windows on one computer, or both ways at the same time. The thinness can be contributed to the individuality of devices and the solo quality of isolation connected with technology. A preference to have it our way when it comes to viewing content promotes thinness in the connecting. We screen off others in the process of watching our personal screen. We can roll out our screen or buy a curved screen, but the basic tenets are the same, a sight priority to a specific target location. And now the sight priority and keyboard functionally of the viewing experience shares space with haptic pens, touchscreens, heads up displays, and augmented reality overlays. Does viewing these screened images make us more or less screened off?

The screen, as part of the production, delivery and display of digital media (and previous non digital media) provides a kind of visual syntax for contemporary digital media tools. Syntax is rooted in language, which is both embodied and cultural. It is rare that our digital media do not have a screen or other projected boundaries. We might touch it as well as view it, but some kind of frame—screen—window is essentially embedded in the technology. The context of technology is understood through the screen and has become a persuasive way of looking at digital media. Media images shift from the film frame to TV and onto computer window, and the viewer shifts from stationary viewer to, in most cases, willing participant. But to be sure, the experience of the frame-screen-window world needs to be further studied because the metaphor lives on. As noted by Merleau-Ponty,

> We must discover the origin of the object at the very centre of our experience; we must describe the emergence of being and we must understand how, paradoxically, there is for us an in-itself. (2000, 71)

The frame-screen-window is that center of the human–technology connection with digital media. We are a part of this world in a way called *a priori*, or the knowing of something without having to think about it. Our body breathes without a conscious decision to breathe. In this way, we inhabit the world and could never think ourselves out of the world for the sake of research interests. I wonder, is the screened world our priority? Are we compelled to look through the window and on to the horizon in the same way we look to the depth and edges of our screened device?

To be part of the world is to do more than take up space; it is to inhabit it (Merleau-Ponty, 2000). We inhabit our digital space by wearing it, viewing it, watching it, sharing it, using it and experiencing it. The centrality of inhabiting occurs through the screen. The computer screen forms a system of planes, foregrounds and backgrounds, seen and unseen, through frames, screens and windows. Merleau-Ponty (1987) would suggest that a body simultaneously opens to participation; a body visible for itself and a self-presence that is an absence from self would be an absent body, a transparent invisibility for oneself. This would explain the absence of the body for someone who has been existentially absorbed by the screen. The participant not only loses him or her self in the software, the body also is lost in the simultaneous opening in self-presencing (16). The environment is not an "other" to us. It is not a collection of things that we encounter. Rather, it is a part of our being. It is the locus of our existence and identity. We cannot and do not exist apart from it. It is through empathic projection that we come to know our environment, understand how we are a part of it and how it is a part of us. This is the bodily mechanism by which we can participate [sic]. (Lakoff and Johnson 1999, 566)

The experience is all encompassing, enveloping and embodying. "Our body is intimately tied to what we walk on, sit on, touch, taste, smell, see, breathe and move within. Our corporeality is part of the corporeality of the world . . . a form of being in the other" (Lakoff and Johnson 1999, 565). This is the human–technology connection. Friedberg notes an explanation from the founder and CEO of the computer design company Autodesk regarding the human/computer connection. He calls them the five successive "user interaction generations: front panel, countertop, terminal, menu, then screen" (221).

The multiscreen advertising firm Collective came up with the catchy slogan, "Life is but Screen." This concept of the screen is foundational when it comes to digital media technologies. The multiple inventions and invitations of the screen push the understanding about the variety of relational elements that occur between the screen and the participant. Screens conjure the metaphors, the interface and the hardware of the user experience. More recently the touch screen has been integrated into design, along with a need for a stylus pen, brush or conductive fiber if our skinned fingers are not available or too chubby for the small-screened face.

While I might prefer to watch a movie on a big screen, I'll view it on my laptop, tablet or smart phone depending on what device I have handy and where I am viewing. I may watch a film on my tablet while sitting in an airport or waiting for my children to finish sports practice, but the viewing experience and all of the aesthetic wonder that goes into a film production cannot be captured through this minimalist viewing experience. I call this "good enough" viewing. It's not ideal but it works. On the other hand, it would seem silly to play a phone game on a big movie screen, but I could. And it is usually better to play a multiplayer game on a big screen, but I don't have to. So there is a screen for every technology and a time for every screen. And in many cases, I use multiple screens at one time—watching a movie on Netflix with my laptop open to do work, while texting a friend. Visual media always has a screen. Digital media's visual precursor, film, used magnified viewers on editing flatbed and upright Moviolas as a way of viewing frames in production. A screen is a necessary component to viewing. One of the fundamental shifts, however, is that digital media has turned the spectator to a participant.

If Merleau-Ponty's (2000) sense of plunging into the things perceived holds true, and the screen is being perceived, then the screen may become an entryway, portal, opening or frame for the user to plunge into in an act of absorption. Sherry Turkle, in *Life on the screen: Identity in the age of the Internet* (1995) said "that the privileged way of knowing can only be through an exploration of surfaces. This makes social knowledge into something that we might navigate much as we explore the Macintosh screen and its multiple layers of files and applications."

The joiner of these frames, screens and windows is the interface. "Interface"—a geometric term for the surface that forms the common boundary between two three-dimensional figures—was deployed to describe the human-computer relation to the computer once the user was literally "facing" the computer (Friedberg 2006, 220). This location gives me a bodily vantage point for perceptual interaction and invites engagement. Engagement is a kind of drawing in and embracing of the mind and body. If you look here you will see "it," whatever "it" is. What does it mean to embrace technology and how might the screen be implicit in the experience? Today's "touch screens" add another dimension to the lived experience of screen use. As Merleau-Ponty explains, "It is not consciousness which touches or feels, but the hand, and the hand is, as Kant says 'an outer brain of man'" (2000, 316). We can touch the screen with our hands or a special glove that conducts a direct current that covers the gap between the technology and the screen. The surface of a touchscreen has a grid of electrodes that spans the gap and completes the circuit. Even technologies like Google Glass do not break the frame-screen-window metaphor because the technology's center for information is a screen that sits out of view and works similarly to a cell phone screen. The glasses frame acts as a track pad, the user gestures, voice commands or uses a finger to scroll and swipe for functionality. Today's "screen capacities" are beginning to morph into a new kind of connection. Perhaps augmented reality will move closer to redefine the metaphoric and physical properties of the screen. A user with augmented reality abilities can add graphics, sounds, haptic feedback and smell to the lifeworld experience on demand, making the world a screen. The user is able to call up computer-generated graphics with the help of a smart phone, laptop or GPS device or glasses, in their perceptual view for use, to access apps that overlays data and information on top of the physical landscape, or personal media contact lens, which makes the world a "canvas" for digital content.

One of the most interesting things about digital media use is that the more screens are negotiated; the more comfortably they are adapted. As noted by philosopher Ed Casey, the longer we reside in places, the more body-like they seem to be. "As we feel more 'at home' in dwelling places, they become places created in our own bodily images" (1993, 120). Bounded regions in space solidify thinking about the abstract virtual places to inhabit. The frame-screen-window, and perhaps now the metaphor of "canvas," which is being used in augmented reality, create perceived three-dimensional space that continually push boundaries as the natural lifeworld and the digital world collapse into one another. This first case study takes broad sweeps to explore a large expanse of ideas that net the perception and metaphor of the screen. Certainly, one becomes interwoven into the other. The perceptions explored here are neither concrete nor exhaustive, but indicative of variations that illustrate human–technology connections. Each of the next eight case studies continues to examine human–technology connection to explore the technological texture.

NOTE

1. A portion of this chapter was presented at the 2014 Society for Phenomenology and Media Conference in Freiburg, Germany and printed in the conference proceedings, titled *Glimpse Phenomenology and Media. Volume 15.*

Chapter Five

Case: Dwelling in Digital Sound

Digital technologies advance a way of relating to the world by plugging in and connecting with devices and content. While screens invite us into a sight priority of images, sound resonates within our being in a different way.

> It is easier for us to shut our eyes than close our ears. It is easier for us to remain untouched and unmoved by what we see than by what we hear; what we see is kept at a distance, but what we hear penetrates our entire body. (Levin 1989, 32)

The connection starts with the way humans listen to sound. Hearing precedes vision and so sounds become part of the initial element of all embodied decision-making (Horowitz, 2013). The ear listens for patterns and interruption of patterns as signs of change in the world. Now more than ever before, human–technology connection is fostered by digital sound. It comes from our devices in speakers big and small in the form of music and other audio content. It touches us and moves us and shapes our emotions. Digital sounds are lifeworld sounds. People are reaching out through their devices to cocreate sound data that clamors through the technological connection. But not all lifeworld sounds are digital. Is there something about hearing natural, unmanipulated, undigitized sounds in the lifeworld that fosters a connectedness in an embodied way that the digitized form of sound does not? Chapter 5 explores the case of digital sound to reveal some overarching themes about the non-neutrality of the human–technology connection.[1]

Frances Dyson, in his book *Sounding New Media: Immersion and Embodiment in the Arts and Culture* (2009), explains audio as "three dimensional, interactive, synesthetic, perceived in the here and now of an embodied space, sound returns to the listener the very same qualities that media mediates: that feeling of being here now, as experiencing oneself as engulfed, enveloped, absorbed, enmeshed, in short, immersed in an environment" (4). And what of this environment that is intertwined with our bodies? Do we plug our ears to hear the sound or plug them to avoid the sounds of the world?

In Homer's poem *The Odyssey*, Circe speaks of the sirens, which sing a sweet song that lures sailors to their death. She tells Odysseus to plug the ears of his oarsmen as they past the place the sirens perch or his men will surely die. The sound of these mermaid-like creatures is so compelling that it overpowers reason. Odysseus wants to hear this song, so he puts wax in the ears of his oarsmen and crew but binds himself to the mast so he can hear the siren's song and live through the experience. This metaphor is a fruitful one for thinking about contemporary digital sound. Is it possible that by engaging in digital sound, by plugging our ears with headphones or earbuds to hear our personally selected digital sound, we are actually closing ourselves off to the sounds of the world? Which environment are the sirens singing in? Could we be lured to destruction through our choice for aural elitism? And finally, does the "we" of the sonorous world bow to the "me" of the digitally programmed and individually selected playlist, podcast or other digital content? This chapter explores the multiple variants of digital sound from a postphenomenological perspective. This case digs deeper to discover the multistability of dwelling in digital sound.

What is the nature of "being with" digital sound? As explained in chapter 1, ontology is our experience of the world through relating in the world. Digital technologies advance a way of relating to the world that is nonrelating in many ways. But the human–technology connection has its own interrelational ontology. By plugging in and clicking on our device, we cut off the sounds of the world. The cry for help, the beeping horn, or the rev of an engine are "drowned" out. Are we succumbing to the siren's song of digital sound? One of the ways we relate is through the use of tools. But if our tools prevent us from relating to each other in fundamental ways, how might the world change? When the priority is the self selected, does our purely compressed and sweetened digital sound playing through individualized conduits from our pre-selected favorites from our previously purchased playlist take priority? As Ihde so eloquently describes:

> Sound permeates and penetrates my bodily being. It is implicated from the highest reaches of my intelligence that embodies itself in language to the most primitive needs of standing upright through the sense of balance that I indirectly know lies in the inner ear. Its bodily involvement comprises the ranges from soothing pleasure to the point of insanity in the continuum of possible sound in music and noise. (2007, 45)

I am again reminded of Odysseus, held fast to the mast of his ship while listening to the siren's death song that penetrates his body while he makes signs to his men to untie him. It seems that our ontological being might be threatened by today's digital sound. When it seems like you're already there, why experience the song live? Are we living a HD life through our audio and our noise-canceling headphones? Does this wax in our ears keep us from danger or move us toward it?

> Our culture is saturated with information, which stubbornly refuses to come alive with understanding . . . We learn to close ourselves off, and we think of our souls and minds no longer as a presence but more in terms of apparatus and function. (O'Donohue 1999, 75)

Abram, in *The Spell of the Sensuous*, writes, "To be engaged in the present moment, is to be in the enveloping field of presence . . . a place which is vibrant and alive . . ." (1996, 204). Is it possible that today's digital sound technology is cutting us off from the present moment? Or is it the other way, that the digital presence is vibrant and alive? Plugging our ears is both a way to both invite sound and remove sound. Both choices invite different possibilities. A "siren" is a nymph that can bring men to death and a warning of danger. Both notions help tease out variations in the human–technology connection for dwelling in digital sound.

Being present is etymologically rooted in the Latin *prae*, meaning before, and *-esse*, meaning to be, suggesting immediacy, promptness and a way of being that takes the lead (Barnhart, 2001). Being present is not always easy. The siren's song might enchant all who hear, but if anyone draws in "too" close, he or she will die. Again we come back to Odysseus, tied fast to the ship mast and present to the sirens song. Presence rests in the idea of being in some space, standing in some specific place as a location and being present as here. Does our experience with digital sound "ring" as true presence?" Can we be here, in an authentic sense, when our listening is somewhere else, disconnected from the sounds of the world? Which kind of listening reminds us of our authentic auditory existence in the world?

TUNING IN

The siren's song of today comes fully available on our digital device. The digital song invites total and undivided listening because of its sharp, complete sound. Like Odysseus on his travels, a listener is beckoned to come closer to new and better quality technology with a crisper purer sound. Can the "sweetness" of their sound be likened to the "sweetened" sound we hear? By definition, *sweetening* is an enhancement of sound that removes impure sounds and layers new ones to build sound using digital software. Might the digitally enhanced din of today's world be the "siren's song?"

It is easier to shut our eyes than our ears during the scary part of a media drama. It takes little effort to turn from the TV during the commercial advertisement but the sound track attached to it sometimes follows us into the other room. Since the birth of radio, programs like Orson Welles's *War of the Worlds* and music movements from popular to fringe have had great impact on culture. Economically a commercial brand attaches a jingle to its product because these sounds are readily heard, moved to the foreground, and identified favorably while shopping. Now the jingle gives way to a few identification notes as the brand becomes branded on listening ears. Levin suggests that "Hearing is intimate, participatory, communicative; we are always affected by what we are given to hear. Vision, by contrast, is endistancing, detached, spatially separate from what gives itself to be seen" (1989, 32). Michael Bull, in his ethnographic work *Sounding Out The City: Personal Stereo and the Management of Everyday Life*, observes that personal stereo users utilize personal audio devices so they do not have to interact with others or the environment, becoming "nonreciprocal," "one step removed" from their surroundings, and "withdraw into themselves" to make their music or other information "all encompassing" (Bull 2000, 25). As Circe tells Odysseus, "Stay clear of the sirens." And as it is told, the siren's flattered him and told him he would be wiser having heard them. The only way of getting past the power of the siren's song was to stop listening. But Odysseus could not do it on his own. It is hard to know what you are missing when you are in the middle of the sound. But the more the sounds of the lifeworld are pushed to the background and a priority for digital sound becomes the precedence, the less careful the listening will be within the lifeworld.

The siren's song of today, the sounds in the media world comes fully available on our devices. The digital song invites total and undivided listening because of its sharp, complete sound. Like Odysseus on his travels, a listener is beckoned to come closer to new and better quality technology with a crisper, purer sound. The contemporary high-tech market leads the consumer to increasingly technologically advanced digital media for viewing *and* listening pleasure. Sound is often heard through amplifiers in a way that allows the sound to physically channel and move matter like water. Better headphones, newer digitization and increased downloading pack a powerful ploy for the consumer dollar to internally channel sound as well. Heidegger, in *Being and Time,* clarifies that "understanding always concerns the whole of being-in-the-world" (1962, 142). Will being-in-the-world in a sonorous way be cancelled out by listening through high-tech devices? Might this beckoning to personalized sound become a closing off of significance in the lifeworld and a shift of increased disconnectedness toward the Other?

SONOROUSNESS

We learn and experience life through the sonorous noises of the world. Sonorous, from the Latin *sonor,* means a noise or a deep loud sound. The notion of full richness within a sound and the pleasantness of a sound also suggest a deepness or resonance in listening. Corey Anton, in *Selfhood and Authenticity* (2001, 91), states, "Sonorousness makes manifest certain configurations of world-experience. It releases and appropriates profiles of the human world" in a way that is "sonorous being-with-others being-toward-world" (108). It seems, then, that part of the manifestation of lifeworld experience might be cut off if world-experience sounds are not heard. Levin, in *The Listening Self: Personal Growth, Social Change and the Close of Metaphysics,* observes that children begin life by listening to the sounds of the world in a sonorous way, "integrating awareness, living well-focused 'in the body'"(1989, 61). But then this highly focused sense of sound goes away as the many sounds of the world move to the background to focus more on specific sounds involved with the foreground of existence. These are the chosen sounds of the digital world. We preprogram or choose many of these specific sounds because of the portability, loadability and availability of our digital media devices. Our "being-with-others-being-toward-world" shifts to being-uploaded-into-the-world as we select and connect to the sounds we wish to hear (108). Kenneth J. Gergen, in *The Saturated Self: Dilemmas of Identity in Contemporary Life,* calls this uploading into the worlds a kind of multiphrenic condition, where we swim in ever-shifting currents of being for a multiplicity of potentials. How might our digital media: human–technology connection influence our "being-with-others-being-toward-world?" (Anton 2001, 108)

NOW HEAR THIS

The attention of today's "eye culture," or focus on the visual, has moved the ear and thoughts about sound and listening to the far reaches of our thinking about media phenomena. Turning attention to sound and listening within the digital world brings the focus to the importance of hearing as part of being. Berendt, in *The Third Ear: On Listening to the World*, suggests, "Hearing is a state of being unmatched by any of the other senses" (1992, 48). Hearing also has profound links to thought, reason and knowledge. Heidegger calls the ear the third eye because of its prominence in thinking (55). So, being-in-the-world is hinged on hearing in the world. How might thinking change because today's listening has become so self selected, so digitally enhanced and self-centered?

Together, the eye and the ear present a "democracy of the senses" (Berendt 1992, 28), an ability to synthesize sound and vision together. But does this consensus continue to be a democracy when senses are dwelling in digital sound? Popup menus have sound effects, memes have music, websites have signature samples and sound tracks highlight digital shorts that auto play on social media. The availability and capitalization of natural and unnatural sounds of the lifeworld are now emphasized more than ever in their digital renderings. Our world is experienced as an "intermingling of lights and sounds, of inner seeing and inner hearing" (33). Some researchers suggest that the media noise might increase aggressive behavior. On the other hand, some teenagers suggest "loud music shuts them off from, modern society's aggressiveness and gives them 'peace'" (79). Might it be possible that the highly individual nature of today's personal and digital sound is also a way of cutting one off from the lifeworld of sound and a move toward a more personal sound of choice, a personal peace? Taylor and Saarinen, in *Imagologies: Media Philosophy*, suggest, "One of the distinctive features of the information age is the proliferation of data whose meaning remains obscure. The more we accumulate the less we have" (Taylor and Saarinen 1994, 8). Is a gathering like this also possible with sound?

RESOUNDING

Personal sound has become an almost constant way to dwell in the digital. What kinds of media sounds are the youth of today bringing down into the body to create home? Ed Casey notes that "bodies build places" (1993, 116), and so then, the body becomes a digital way station for today's youth. For a place to become a dwelling place, the structure must allow for repeat return. Music technology, loaded onto an mp3 device or mobile phone connected to earbuds or headphones, creates a feedback loop of digital dwelling structure as a repeatable configuration. Sounds, songs, messages and a variety of other digital content are repeatable. Additionally, there is that "connected" feeling of familiarity because each digital device is personalized to a specific taste. A digital-music player has specifically selected and downloaded music, the phone has specific numbers on speed dial with a favorite ring tone and the digital-media device has all of the necessary lifeworld information to plug in and connect. This is living in the familiar. If the two criteria for a "full fledged dwelling place" (116) have been met, might our technology be called our digital dwelling place?

Video games with audio tracks of digitally altered sound effects are available for most digital-media devices. Music players bridge the gap between the ear and the sound to create a "seamlessly pure" listening experience that becomes a habitat, that becomes dwelling as perceptual listening develops into an all-encompassing sensory experience. If digital makes it better, more crisp and more pure, then the media sounds—whether positive or negative for young ears—will be heard in a crisp and pure way.

FILTERED OUT

When audio professionals discuss the creation of digital sound, their concerns include the notion of *pure* sound. Is it important for recorded sound to most closely replicate the original sound when it occurs or just to make a sound that appeals to the listener? Do the recording devices take away from the listening or can the listener bypass the faults of the recording device?

The notion of pure sound, with regard to recording, can be ambiguous. The etymological roots of *pure* suggest a Latin root meaning lean, chaste, or unmixed (Barnhart 2001, 865). Middle Irish and Welsh roots also imply something fresh and new or sifted. Manipulated sound as well as unmanipulated sound can be seen as pure. Real sounds, however, suggest etymological roots that mean actual, true, genuine or authentic (891). This distinction becomes important when sifting through ideas about the impact of digital sound.

The multistability of digital sound presents many different structures. I wanted to know what shows itself when audio professionals were asked if they thought that digital sound was different from sound prior to the digital age. Several meaningful variants emerged. In an online discussion among audio professionals in a peer-to-peer creative digital community called creativecow.net, I asked the question, "How do you think digital has changed the notion of sound—for media, for entertainment, for personal communication? Sometimes I think that what I hear in real life is not close to what the manufactured sound is like—yet it is described as being better quality, just like the original." Four site-registered audio professionals, Brian B., Peter P., Ty F. and Rob F. responded to my query and reveal interesting variations about the nature of digitally manipulated sound from the perspective of those who create today's media. The perspectives, multistabilities, and shifts in ideas about differences between a real duplication of an intended sound and a fabrication of sound reveal new ideas about human–technology connection.

REAL/PURE

These audio professionals were wrestling with the ideas of creating new sound versus replicating older sounds better, and how to rectify real versus pure sound. The idea of replicating and copying sound also comes into play, and words like "pure," "real," "identical," "reproduced" and "cleanly produced" are used to describe the variations on whether sound is pure or real. Brian B. suggests:

> that audio professionals in the past were always trying to make a recording sound like the real thing or make it sound like something you've never heard before. I think currently with digital that people are creating sounds and are not trying to duplicate nature; they are trying to create a totally new creation . . .

Audio professionals wield a fairly powerful hold over what a particular audio bit ends up sounding like. Like early Foley artists who created sounds for film with a variety of different objects, today's sound engineers manipulate the audio and choose what is real and what is Memorex. Most sounds are digitized into a computer and manipulated through a variety of mixing programs. Brian B. suggests that a real sound has taken the back seat to a pure fabricated sound. So then, if fabrication is now easier through digital, and a priority for creative artists, is there any obligation to the authentic? Are there ethics for real audio? Brian B. adds: "But for some reason gunshots are different. They always have to be bigger sounding and bassier than the real thing." What might propel an audio professional to feel that a gunshot sound needs to be bigger than the real thing? Where does this sound mythology come from?

Rob F. suggests that the listener has been willing in the past, to forgo some added recorded noise and buy into the believability of the sound. Maybe today's listener is willing to buy into the sound, because it is so pure, whether it is hyped up beyond the real or not.

> You can find vintage advertisements that state words to the effect that "the Edison Phonograph reproduces sound so clear that you'll think the artist is singing right in the room with you!" No one could POSSIBLY think that an old wax cylinder with its horrendous surface noise, severely limited frequency response, and wow and flutter could sound "REAL" . . . but it was so engrossing that folks could ignore the artifice and just concentrate on the "moment of enjoyment."

So it seems that there are two separate issues here. One notion involves the idea of whether a digital sound is true to its original and the other is whether it continues to be a clear duplicate of the way the audio professional originally intended it, enough to be enjoyable to the user. Whether it is a real duplication of an intended sound or a fabrication that sounds like the professional wanted it to sound might not be the issue at all. One choice is based on technical apparatus and the other is based on the creative license for enjoyability and access for the consumer. One is about the creation of a pure sound from the ground up, and the other is about pure duplication of a real sound.

Case: Dwelling in Digital Sound 71

Either way, the listener makes his or her own meaning of the noise. Rob F. suggests:

> What is the MOST important thing to consider (and if you are too young to remember otherwise, you'll just have to trust me) is that until modern digital recordings . . . it simply was not POSSIBLE to make a recording (and subsequent copies) and have the playback sound virtually IDENTICAL to the source in the monitors. THAT is very significant. So whether it sounds like "life" or not . . . it DOES sound (virtually) exactly the way the studio engineers heard it when it was mixed and recorded.

The exact duplication of a sound, any sound, through most of the copying and manipulation, is the siren song of digital. The "how" has become more important than the "what." The ability to create pure sound beckons the audio professional to push the sound to its limits instead of duplicating the sound as it was originally heard. Ty F. notes, "The copy medium is stunningly close to the original." A closeness or inability to distinguish between the sound of origination and the final destination copy that ends up in the hands of the consumer is a key focus for audio professionals because an ability to duplicate was very difficult to achieve before digital. Peter P. adds:

> I mean, when you are the one in the control room mixing the session . . . and can then REPRODUCE EXACTLY the same audio track—sans clicks, hiss, dropouts, inner-groove distortion . . . How can there be any doubt that digital is a more PURE medium? I remember when we had LP's made from our master tapes . . . The audio quality DROPPED by a horrendous amount (even if the master was "specially" EQ'ed). If the sound is FLAVORED by the analog process, why would adding FLAVOR be a "good thing"? Consider wine. Glass bottles would be the "neutral" medium for wines. Digital is the "neutral" (at least MORE neutral than analog) medium for audio.

The etymology of *neutral* stems from the Middle French idea of a compound of contrasting elements or a gender that are neither masculine nor feminine. Neutrality stems from the root *no* and *either* or *neither* (Barnhart 2001, 702). So, is digital just a neutral medium or does it bring along its own, unintended artifacts? Is it a compound of contrasting elements? Peter P. adds,

> Let's face it . . . until we can sample at infinite sample rates, we must be leaving behind some information, so digital recordings add their own "flavor" just like vinyl does.

What flavor might this new digital sound create for the listener? When the real and the pure are indistinguishable, what is left for the listener to hear? And when the lifeworld is made up of uploadably fabricated sounds tailored to individual liking, how can the real world sounds compete? Brian B. notes,

"Children and adults nowadays get too 'lost' into media. They believe it is real, when in most cases it is very far from 'real.'" Brian B. warns about the difference between pure and real when he observes that people can get lost or "be deprived of"(Barnhart 2001, 610) the world because they become so engrossed in the media, a "far from real" world.

This distinction in the digital media, human–technology construct becomes pivotal in thinking about digital dwelling because it suggests that the media world might be away from home instead of being one's home. Can somebody be lost in the media world and still be in a digital dwelling place designed by media apparatus? Do notions like pure, real and reproduced make a difference for our human–technology connection?

Maybe the spirit, or "breath of life," of digital dwelling is an important consideration. This root notion of spirit brings forward the kind of identifying properties that today's world exhibits toward digital media devices. This vital link or breath is connected and entwined to create the dwelling place. Shutting off the lifeworld sounds of media in a turn toward self-selected sound is a breath of fresh air for teens and young adults. Loud music coming through bedroom doors and from automobile windows used to illustrate the growing pains of youth culture for all to hear. Today, the sounds of this digitally created world are of the self and not of the lifeworld. What will a world of selves, carrying their media homes on their backs like snails, move the world toward as a future? Are these dwelling places self-limiting or peacekeeping? Are these homes linked to the net in a communicating way or a disconnecting way?

Possibly today's youth culture is trying to hold onto and create a digital dwelling place in the spirit of home by creating a dwelling place of peace away from the world of technology. They are using the tools they know best, the digital tools they grew up with and plugged into at a very early age, because now they are coping with too much information in the world. Digital dwelling is a way of "coping in a culture, overcome by information generated by technology, [that] tries to employ technology itself as a means of providing clear direction and humane purpose" (Postman 1992, 72). Furthermore, "as humanity rushes into an age of ever-increasing technological sophistication, and at the same time becomes more aware that technology unconnected to nature's laws is suicidal, it becomes apparent that traditional wisdom holds keys to restoring sanity and balance in our lives" (Swan 1990, 75). This traditional wisdom centers on an understanding of authentic and real, not pure, listening in the lifeworld.

TOWARD DWELLING

Raising awareness about the sounds of the pure versus the sounds of the real, those heard in nature, in real life, can be the key to bring young people back to interpersonal communication, back to "sonorous being-with-others-being-toward-world" (Anton 2001, 108). Levin explains:

> In listening to the sounds of nature, listening to the music of sounds, and listening to the speech of others, we learn, we grow, we help others to learn and grow, and we realize that hearing is a gift to be valued and enjoyed . . . But how capable are we, each one, of becoming, in Rilke's words, 'a being with no shell, open to pain . . . shaken by every sound.' (1989, 89)

When we are emplaced, we are in a comfortable place that we know, and "threats to this emplacement are also threats to our entire well being" (307). There is pain in breaking out from the digital shell. "It is less a question of what gives 'true peace and contentment' than of comparative priority" (313). The elevation of place as a consideration reveals much about digital dwelling. This lived, phenomenological experience of noticing how it feels to be connected in a sonorous way helps unravel the understanding of the balance between the inward place that shuts out the world through the technology and the connectedness to the outside place. When young people inhabit their digital shells they are not displaced as much as placed somewhere else. They are not home, specifically, but celebrating the spirit of home through the technology. "To get into the spirit of a place is to enter into what makes the place such a special spot, into what is concentrated there like a fully saturated color" (Casey 1997, 314) that sweeps the self and others into its folds in an encompassing way. This being-in-place in the lifeworld does not have to be insular. This experience combines the self and others, the noise and sound, the mind and body, into a totality of dwelling and being. Home only feels good after you've journeyed away and found yourself in a new place and are ready to go home again. The intermingling of the personal and the collective life makes getting lost mean something, because people who are lost can move to find their place again.

What is this close link between the self and the digital dwelling place? The satisfaction of digital dwelling resides in keeping in touch with the world. The secret of the digital dwelling place is situated in the places of the world, not in the mind of the self. This placing and replacing of one's self on the digital highway with a variety of ways to reach out and touch someone needs to be daily rooted in the real, in the authentic, to keep from getting lost in digital media. Careful mindfulness toward a knowing, experiencing the places on the journey alters and turns outward instead of inward. The body-in-technology is still a dwelling place, but it is also an authentic place of spirit that will "reconcile itself with itself, to recognize oneself in other being" (Gadamer 2000, 13). By acquiring the skills, the capacity of digital dwelling, today's youth gain a sense of and a home for themselves. What might be seen as alienation is more of a return to oneself to understand the world and negotiate it. We hear ourselves through the digital cable as we connect.

Sound's role in this new way of understanding digital dwelling is simple but profound. When is it clear and when is it real, and what is the balance? The celebrated film editor and sound expert Walter Murch says, "It's always a balance for me, between something being authentic, and celebrating that authenticity, and yet at the same time trying to push the sound into other metaphoric areas" (Ondaatje 2002, 120). In this digital age of computer manipulation, do the real sounds in the real world matter? It seems that they must, because even when we sleep, when all other senses are shut off, we can still hear. Berendt notes that "we can hear before we enter this world, throughout our lives, and even in the hour of death when all our other senses fail us, which demonstrates that hearing is a state of being unmatched by any of the other senses. . . .

If hearing is essential, what we hear, what we connect with in aural habit, is fundamentally significant. The listening self hears the sonorous world. Does digital dwelling foster being in the sonorous world? Heidegger (1962) explains,

> suggesting having to do with something, producing something, attending to something and looking after it, making use of something, giving something up and letting it go, undertaking, accomplishing, evincing, interrogating, considering, discussing, determining . . . all these ways of Being-in have *concern* as their kind of Being—the being of a possible way of Being-in-the-world. (57)

It seems important, then, to think through one's being-in-the-world in the age of digital dwelling. The more the sounds of the lifeworld are pushed to the background and a priority for digital sound becomes the precedence, the less careful the listening within the lifeworld's non-digital sounds becomes. Heidegger calls careful and intentional listening within the lifeworld a "listening to understand" (Heidegger 1953, 164). And Levin writes: "If we listen well to ourselves, we can hear within our embodiment resonances and echoes that confirm the interconnectedness of all beings . . . gathering us together for the making of a more thoughtful history" (1989, 272). Does dwelling depend on the purity or realness of digital sound or the perception or realness? Had Odysseus himself experienced dwelling when he heard the siren's song or did his men experienced authentic dwelling by plugging their ears? Do embodied resonances, whether digital or live, lead to dwelling? Understanding these multistabilities of digital sound will bring forward a more thoughtful digital history and a clearer understanding of human–technology connection. Listening to understand is the right frame of thinking as we explore a few more cases in digital sound, namely earbuds, iPods, and dubstep.

NOTE

1. An initial exploration of this case was shared in a paper given at the Society of Existential Theory and Culture Conference, Vancouver, British Columbia, Canada on 6/5/09.

Chapter Six

Case: Earbud Embodiment

Connecting to personal sound has soaked into every facet of the lifeworld. As explored in chapter 5, dwelling in digital sound is a profound human-technology connection. One way to explore a more specific human-technology experience is to focus on one aspect of digital sound like music listening with earbuds. This pastime-turned-ritual is both a macroperceptual and microperceptual experience, but the digital listening technology of contemporary society has brought a shift in embodied experience, from carrying to wearing to literally plugging in.

The deep and clear channel of personalized music and other digital content brings joy to those who plug in, but this non-neutral technology has residue. At one time people walked to the beat of the lifeworld rhythms. While not a chosen soundtrack, beeping horns, chiming clocks and radio station identifications marked the hours in the day and signaled approximate time changes, like noon or the start of rush hour. Train horns and bus pistons signaled the end of school and the end of the line. Sounds marked time and illustrated experience. You either absorbed it in or shut it out. Digital sounds are the new lifeworld sounds we live by, but in many cases, only we can hear the playlist. What is the residue from picking and choosing a lifeworld soundscape of one's own? When the sounds of the lifeworld are pushed to the background and a priority for personally selected and digitally programmed sound becomes the priority, listening and hearing change. Heidegger calls careful and intentional listening within the lifeworld, a "listening to understand" (1953, 164). But what of our digital lifeworld sounds? The podcast we choose or the playlist we load? Are chosen lifeworld sounds also a listening to understand? Or does the choosing negate the understanding? Levin writes: "If we listen well to ourselves, we can hear within our embodiment resonances and echoes that confirm the interconnectedness of all beings . . . gathering us together for the making of a more thoughtful history" (1999, 272). What happens when digitally programmed and channeled sounds become the preferred lifeworld soundscape? Will we still hear our body resonate in interconnectedness? What will that gathering look like?

Levin observes that children begin life by listening to the sounds of the world in a sonorous way, "integrating awareness, living well-focused 'in the body'" (1989, 61). But then this highly focused sense of sound goes away as the many sounds of the world move to the background to focus on the foreground of existence. These are the earbudded sounds of the digital world. We preprogram or choose many of these specific sounds because they are in portable digital files, easily downloaded onto our device, and available any time we choose. And choosing one sound denies another. Just by choosing to listen to music, we decide that other sounds are less important.

Through earbuds, we link to all kinds of content, like music, podcasts, conversation and a variety of other digital programming. The bringing of one's personal area into public spaces has been occurring for some time, with cell phones, laptops and other smaller high-tech things, but the earbuds provide something at the same time that they limit something else, namely environmental lifeworld sounds we use as information to negotiate our embodied day. People have died because a car or train hit them because they could not hear the world around them. Because of this, marathon and triathlon runners who run sanctioned races are not allowed to wear a music player during competition or can only legally wear their music in one ear for safety reasons.

The purpose of earbuds is to transport sounds, and these sounds can affect us profoundly. Ihde notes:

> Sound permeates and penetrates my bodily being. It is implicated from the highest reaches of my intelligence that embodies itself in language to the most primitive needs of standing upright through the sense of balance that I indirectly know lies in the inner ear. Its bodily involvement comprises the ranges from soothing pleasure to the point of insanity in the continuum of possible sound in music and noise (2007, 45)

Postphenomenology recognizes both the deconstructive and yet structural in the lifeworld. The perceptual embodied experience absorbs and attends to both.

Creating our own sound environment is entertaining, it passes the time, it helps connect us with parts of our culture and history and it may help us feel safe as we control our sensory environment where we can. Kenneth J. Gergen, in *The Saturated Self: Dilemmas of Identity in Contemporary Life,* calls this uploading into the worlds a kind of multiphrenic condition, where we swim in ever-shifting currents of being for a multiplicity of potentials. Our "being-with-others-being-toward-world" shifts to being-with-others-being-toward-self "as we select and connect to the sounds we wish to hear, while still operating bodily in the lifeworld" (Gergen 1991, 108). In what ways might this human-technology connection impact our "being-with-others-being-toward-world"? (Anton 2001, 108) This question hangs within the inquiry of our earbudded embodiment.

Humans used to focus on the listening by intellectually pushing away unwanted, background or ancillary sounds by controlling the environment. Going to a quiet place, studying in a library or finding a room of one's own controlled the experience. In contemporary society, the space stays the same but the listening is altered by technology. Explains Merleau-Ponty, "We think we know perfectly well what 'seeing,' 'hearing,' 'sensing' are, because perception has long provided us with objects which are coloured or which emit sounds. When we try to analyze it, we transpose these objects into consciousness. We commit what psychologists call 'the experience error,' which means that what we know to be in things themselves we immediately take as being in our consciousness of them. We make perceptions out of things perceived . . . we are caught up in the world and we do not succeed in extricating ourselves from it in order to achieve consciousness of the world" (2000, 5). But digital technology has altered our way of becoming caught up in the world.

Today digital-media users focus on one specific accessory to more directly channel sound right into the ear and toward the eardrum. This accessory is called a pair of earbuds. These tiny speakers that hook into the ears come with most digital media technology, but can be purchased separately almost everywhere, including convenience stores. The focus on this case study is the specific micro- and macro-perceptional elements revealed when using earbuds. The aim is to explore the ways earbuds are a non-neutral technology.[1]

The current ear-wearing sound device variation, called earbuds, are typical instruments in today's digital-media device market. Ear covering sound devices have been around for a while now, but they were not always so *bud* like. One of the earliest headphone type variations, a heavy single cuff called an electrophone, rested on a telephone operator's shoulder to aid in hearing connections. When headphones entered the consumer market this variation was bulky and mostly used for musicians, radio work, and professional sound recording. In the early days of sound technology development, speakers were also fairly large in size. But in the 1960s a small, personal pocket device, called the transistor radio, changed that. This small battery-powered radio sported built-in speakers that amplified sounds from radio station airwaves through analog technology. While larger radios and both attached and unattached speakers were still popular for home stereo use, portable varieties took to the streets, which popularized the portable sound experience. In the 1970s a larger portable radio, called a boom box, featured larger loud speakers, and the idea of portable sound became cemented in youth culture. In the

1980s the personal cassette player, which was barely larger than a deck of cards, entered the market and became popular with teenagers and athletes looking for a "mix-tape" for workouts. These devices often came with a small spongy ear covering headphone variation that personalized the sound experience and made the music easier to hear in a competing lifeworld of sounds. The smaller, lighter spongy-ended headphones quickly became popular, along with many brands of personal radios, cassette players and later, portable CD players.

The mp3 players of the early 2000s, fueled by the newly released iTunes software, brought the headphone quickly forward as a necessary device, because mp3 players did not come with external speakers. Some kind of listening technology was required. Portable became personal as the surround sounds shrank to internal proportions. While speakers that project out into the world compete for sounds with the lifeworld, headphones, in ways big and small, work to remove competing sounds to isolate specific ones. The newer headphone variation, earbuds, are a discreet technology but not an invisible one. They can be hidden in a hood or wound into hair but usually can be seen if someone is specifically looking for them. Wearing earbuds are a nonverbal cue for many things, but the most evident one is the wish to change unwanted sounds with preferred ones. Can I hear music from my player without the earbuds? Not unless I plug in a different translatory tool, like a speaker. The miniature Bluetooth or wired speakers called earbuds often have cushioning around them, both for comfort and functionally, to block outside sounds from coming in and keep the sound from the player plugged directly into the ear canals. They are designed to listen without disturbing others, but also to exclude sounds from the outside. We process sound differently from sight. Sometimes, "It is easier for us to shut our eyes than close our ears. It is easier for us to remain untouched and unmoved by what we see than by what we hear; what we see is kept at a distance, but what we hear penetrates our entire body" (Levin 1989, 32). A personalized playlist, facilitated by earbuds, individualizes one's life with a soundscape of prerecorded and preapproved information, which has a variety of lifeworld effects.

Headphones have become such a popular and necessary part of our sound technology experience that it has proliferated the idea called headphone culture. And most interestingly, the headphone technology does nothing specific on its own except facilitate sound between the human and the device. Earbud technology also facilitates the ability to jack to another device. This connective technology can be used in several different ways and has increasingly gained status as younger generations re-embrace a new brand of headphone culture. In his article "Headphone—Headset—Jetset," Sean Nye explains that headphone culture focuses on the idea of private listening but also explores an interesting tension, or dichotomy, between public and private space listening within different cultures. He notes, "The dialectic is reflected in the three primary musical uses of headphones . . . music consumption, studio production/sound mixing and DJ performance" (2011, 67). Earbud embodiment specifically deals with music consumption. The genesis from those early headphones used in sound and recording to the earbuds of today have given these devices a solid footing in the world of private listening.

After purchasing my first pair of earbuds, I was almost immediately reminded of the "seashell ear-thimbles" in Ray Bradbury's 1953 novel *Fahrenheit 451*. In the book, Montag says of his wife, "And in her ears the little Seashells, the thimble radio, stamped tight, and an electronic ocean of sound, of music and talk and music and talk coming in, coming in on the shore of her unsleeping mind. The room was indeed empty. Every night the waves came in and bore her off on their great tides of sound, floating her, wide-eyed, toward morning" (52). Sound waves, both figuratively and literally, transport their users into a personalized soundscape of varying degrees almost every day. This case study explores a similar personalized soundscape.

Embodiment is a crucial part of the earbud experience because earbuds are wrapped around the body and plugged into the body. Whether tethered or wireless, in-ear devices are the most popular way to listen to digital sound on the go. Motility is gained once the earbuds fit the ears and the activity level of the user. Merleau-Ponty would explain this human-technology connection as "thinking the mortal body in the light of Being" (2000, 62). When earbuds fall out or hurt the ear, the technology is no longer transparent. But when ears are snuggly budded with the small plastic nodes, a personal playlist is at the ready to privatize sound. Transparency comes from the user experience of wearing them and from the spectator view of the embodiment of them. David Michael Levin, in *The Listening Self: Personal Growth, Social Change and the Close of Metaphysics,* makes clear that "even one and the same thing can present itself in numerous different ways: as present, as near and close, as past, as distant, as absent, as a perception . . . as a sound" (1985, 131). And Ihde explains, "What is present and said always carries with it what is present and unsaid" (2007, 149). So I wonder, what is present and unsaid about earbud embodiment?

WIRED FOR SOUND

How is it to be an earbudded body in the world? What is the realness of the experience? Earbuds are non-neutral and they are becoming deeply imbedded in daily life. They give a distinct shape to the lifeworld, a different shape than their predecessors, the headphones. With earbuds, the bodily focus shifts from the head to the ears. Wearing earbuds is about privacy and isolation—keeping sound in and sound out. Etymologically, *isolate* has historical roots meaning, "making an island." The idea of isolating oneself to the outside world in order to focus on specific sounds that go directly into the ear without distraction is the original use for this technology. The user becomes an island, land surrounded by sea, when the earbuds are in. Debates within the public sphere center around the socially and culturally deemed appropriate and inappropriate times to wear earbuds. Is there an earbud "etiquette?" I am bodily directed in an environing world, full of cultural social meaning and construction. Think about the set of variants for reading an image of someone wearing earbuds. Is this person approachable or unapproachable? Can they be bothered? Does the earbuddedness suggest anything specific? What is the "speaking" of the image of the earbud user?

The very act of putting in earbuds sends specific social and communicative cues that a user is giving auditory priority to one kind of perception and choosing to (mostly) exclude another. Wearing earbuds means different things to different people. While some people might think it is rude to "plug one's ears," others think it is completely normal. Some might feel that NOT wearing earbuds on the train, or while walking around town, is odd. What are the social-cultural meanings of putting pieces of plastic in one's ears and walking around? And what are the implications of the wearing? The very image of the earbud-wearing body can be "read" in multiple ways. The reaction is both actional-perceptual and cultural as I am being seen by another as experienced also by myself. Earbuds, when plugged in to a digital media device, make one world available to me—a digitally programmable soundscape of my own.

What do earbuds imply? First, they imply that I have ears and can hear. They are portable and small, so they can move with my body, they rarely restrict my body, and they can go somewhat undetected until someone can see my face. Are they designed to block out all sounds? Possibly, if they are in tightly enough, but outside sounds, especially loud ones bleed through. Earbuds trade smallness and portability for larger noise canceling ear cuffs. They are specifically designed to be inconspicuous and light. Some people thread earbuds into their hair, clothing and backpacks. Early twenty-first–century technology has gained in popularity and prominence along with the Apple product line. Now cell phones and tablets also carry music. Digital radio station app options grow. Personal sound is here to stay, and with it, the earbuds.

The postphenomenological approach reveals the multiple ways technology, embodiment and communication interweave. Studying and analyzing this technology's use provides, notes Ihde, "insight into the very structure of multistability, an insight that then can guide our subsequent awareness, such that we might well expect both more possibilities and . . . multiplicity of profiles" (Ihde 2007, 201). Thinking this through can help develop a way of constituting ourselves in the mediated world. Listening through technology invites a specific kind of sound and cuts off another kind of sound. The multistability of earbud use helps clarify new and different ways of thinking about culturally embedded images of choosing to direct one's sound directly to ones ear, via wires. Postphenomenological investigation also can provide evidence for further theorizing and, to shift thinking and make the unseen seen, or the unheard heard. Proposes Ihde, "to both 'see' in an embodied position, and to 'read,' in an apparent position, and to be able to 'hermeneutically' transpose between the two positions is part of what it means to perceive in the *now postmodern world*." Our perspectives are multiple, refractured, and compound (87).

VARIENTS AND INVERSIONS

As previously discussed, technologies have both intended and unintended consequences and a variety of trajectories. Once obsolete, many of our most-loved technologies become museum displays or junk drawer castaways more quickly than intended. Use patterns change, hacks are discovered and popularized, and usage is pushed in different directions once users get their hands on a device. For instance, in a reverse of sound fortunes, earbuds could be used as a rudimentary microphone by attaching them to paper cups, megaphone style. This inversion allows technology designed to go into the body the opportunity to be amplified with a bit of plastic or megaphone-styled material. Postphenomenological variants and inversions can uncover the multiple paths and use patterns of earbud embodiment. The intertwining with this technology says much about our human-technology connection. A micro-perceptual lens highlights the sensory experience and its call to the ears and down into the body. Earbuds are specifically designed to nest in the ear. They are designed to take up a specific bodily position and both fundamentally and paradoxically open and close channels of auditory information about the world. A body wearing earbuds situates itself toward the world in a way that accepts that specific sounds can and should be closed off. Earbuds prioritize sounds in a different way than my brain does, when it notes a person walking toward me from a distance, the rate of that person's movement, and the loudness that I might choose to cut off or dull by putting my hands over my ears. Macro-perception, conversely, discovers that this specific perception of one channel of sound, to plug off another, unwanted sound, can be socially acceptable and completely appropriate. Contextualized, there is a cultural shift of communication based on earbud use in the lifeworld. What is the knowledge held in the earbud-wearing person and the knowledge held in those who read the embodied experience of the user, or the graphic representation of the earbud embodiment experience by technology brands? There are earbuds, as a technology, and the meaning of earbuds, which are multistable in meaning. Together, the multiple ways of relating and communicating in today's technological texture can be examined. The next section analyzes five different variants explained as wired for music, noise canceling, accessorizing plug, controlled sound and avoiding connection. All of these variants are micro-perceptially embodied ones because the body is implicit to the earbud-wearing experience. Each variation also carries macro-perceptual effects. The way we relate to and with one another is based on social and cultural norms. Earbuds have altered that lifeworld experience in a variety of ways. They create a conduit for music and other programming and content, cancel out unwanted sound, act as an accessory, work to control an environment, and allow us to plug in to our choice sound and tune out unwanted sound.

Wired for Music

The marketing for earbuds was initially linked to music players. The silhouette of a dancer holding a small box with a string leading to the ears became an early graphic for the Apple iPod brand. Previous mp3 players used headphones and earbuds too but the earbuds that were emphasized in the Apple iPod brand helped to catapult their use in mainstream society and embedded them in pop culture. The graphic soon became an iconic branding image for all things iPod. The earbuds and player were always prominently featured in white against a black silhouette of a body. The focus was to hold the iPod in hand and dance to the music, hair flying and clothing swinging in rhythm, experiencing fun. In the graphic, the iPod was always shown. The status of the new device was part of its allure. The mystique of the brand helped foster this. An earbudded user body became the mediator of the lifeworld, that place where essence and existence meet and become bound together in existence (Merleau-Ponty, 2000). Understanding the idea of the flesh can help to explore the idea of habit and unconscious awareness. The earbuds become part of our bodies as "habit expresses our power of dilating our being-in-the-world" (Merleau-Ponty 2000, 143). We become one with the technology. The existential idea of motility, spontaneous movement, shows that the earbuds are well in hand, if not well in ears. The perceptual focus between sound awareness and sound transparency becomes a back and forth play of embodied and disembodied perception. The auditory world is part of our embodied word, our field of being, our flesh.

> Flesh is a notion which finally makes it possible for us to articulate the human body with respect to its ontological dimensionality: its inherence in the field of Being as a whole, and the destiny, or ideally, for which this inherence claims us . . . Merleau-Ponty's notion articulates a corporeal schema which roots the human body, as a local opening and clearing, in the multi-dimensional field of Being, for it articulates the embodiment-character of our responsiveness and

elicits its potential for development on the basis of our initial, most primordial sense-of-being-in-the-world. (Lakoff and Johnson 1999, 62)

If my apparatus is digital and I do not want others to hear what I am listening to, I use earbud technology. If I create this kind of earbud embodiment, one that articulates a corporeal schema that is inward turning, does my primordial sense of being-in-the-world change? Most mobile phones, music devices, tablets, computers and gaming devices have a mini jack for a universal headphone port. The technology comes ready and willing to individualize. The choice is not in privacy but in sharing. While larger headphones have become a fashion accessory and status symbol as well as an audio accessory, capturing music within the ear cavity is the essential job of the small and portable earbuds. Earbuds soon became part of everyday wearing out and about in the lifeworld. In school, on the street, at the local gym and in the subways, users had earbuds dangling from their ears and plugged in to something, somewhere. The variety of devices with standard earbud jacks proliferated and device use shifted from the music player device to mobile phones, personal gaming systems and even to an unplugged location in one's pocket to prevent communication with others in the lifeworld.

Noise Canceling

Earbuds were always about isolation of outside noise for focus toward the "deviced sound," but another way of using earbuds is as a cultural sign stating that the user does not wish to be bothered or engaged in the everyday sounds of the lifeworld. Avoiding eye contact with earbuds in, users young and old found ways to avoid communicating with others. People tended not to interrupt someone's music listening. Earbud/headphone wearing meant that the user was concentrating before the big race, studying, talking on a mobile phone or "chilling" to music. The technology allowed the earbudded user to avoid conversation or attention to the world around them and promote individual separation and isolation. Closing off lifeworld sounds also closed off lifeworld connection to small talk, inquiry and interest. But in a world where stranger danger becomes the law of the land for little ones, the idea of talking to a stranger on a subway ride becomes unthinkable. Closing off the lifeworld through earbud embodiment solidifies and discourages interaction with those we don't know, and those we do know but do not want to communicate with.

Accessory

Wearing earbuds is an often-transparent embodied experience. So much so that earbuds have become an acceptable apparatus in almost every lifeworld setting. Even when not in use, they dangle around the neck or through hair, ready for use at a moments notice. Just a short time ago it would have been considered rude on many levels to plug one's ears and choose not to listen to world sounds. Today, plugging one's ears gets barely a glance. When earbuds first became popular, they were seen as a "fringe gadget," something that could be useful but not necessary to complete a task like music listening. Not so today. They have become an essential instrument for getting through the day. Plugs in ears join rings on ears as fashion accessories and everyday objects. At one time devices came with speakers for hearing music, but portable personal technology emphasizes a small size at the expense of speaker space on the device. In many cases the sound from these devices cannot be heard at all without earbuds. Bluetooth speakers have added to the repertoire of available amplification of sound, but private listening still rules when it comes to digital media devices. The word *earbud* makes a way for this apparatus—buds, tiny buttons that can bloom the sound like large speakers, for the ears. The more plugged-in the user is, the more tuned in or tuned out (or both) the user becomes.

Avoiding Connection

In the increasing frenetic, media-saturated lifeworld, users need to set limits and boundaries to control sound in the lifeworld. Not only do earbuds cancel, they control sound. The world is loud and becoming even louder. Urban sounds concentrate in crowded neighborhoods where people, traffic and culture comingle. Suburban neighborhoods focus sounds in planned playgrounds, pool complexes, skateboard parks and community gardens. Rural sounds compete with wind turbines and highways. The earbuds go in because users like to control the sounds in their individual world. In home life and the working world, in the city and the country, earbuds create a controlled and "handpicked" soundbed to "do life" in. The technology becomes the cohort through the human-technology connection. Sometimes users keep one earbud pulled out and dangling to let lifeworld sounds in, and other times they are pinched in tightly. And sometimes the earbuds are switched out for noise canceling headphones to promote silence. The parent listens to uplifting music while the children are fighting in the background; the commuter filters the clatter of public transportation and the sounds of the stranger's arms length away. Children listen to the soundtrack on the game through earbuds so they do not distract the adults around them. Athletes create playlists that keep the focus on the heart rate and the cadence of the exercise. Those who do not like the environment they are in, pick a soundscape closer to "home." Earbuds offer control of a main sense within the environment, hearing.

Plug In to Tune Out

The most widely held perception of music wearing is that a person wearing earbuds is listening to something coming through the tiny personal speakers hooked to a personal digital player attached to the body through straps, cords or clips. However, earbuds can also be used for other reasons not associated with the gathering of music or programming. Some wearers actually turn the music off but continue to wear the plastic coverings in their ears to avoid interpersonal communication. The current widely held notion that ears "plugged in" to the earbuds to listen to something specific may be false. The sign of earbuds in means "Do not bother me." An entire realm of communication has been cut off in many ways because humans do not communicate to pass the time in the same ways that they used to. On the bus to camp, kids play individual games and listen to music through their earbuds. Waiting for public transit on a city campus, students are earbudded and closed off from quippy conversation and getting-to-know-you camaraderie. All because their ears are closed off to these possibilities. A voice that is heard tells much, but a voice unheard also tells much.

WORLD SOUNDING

Each of these variants affects the human-technology connection in different ways. And all have a net effect on the way we read, and hear, the world. The variants of wired for music, noise canceling, accessorizing plug, controlled sound, and avoiding connection have wide-reaching lifeworld implications. The idea of sonorousness, explains Corey Anton in *Selfhood and Authenticity* (2001, 91) "manifest[s] certain configurations of world-experience. It releases and appropriates profiles of the human world" in a way that is "sonorous being-with-others being-toward-world" (108). It seems that part of the manifestation of lifeworld experience might be cut off if world-experience sounds are not heard. In other words, if a person calls out and no one hears the call because everyone in the lifeworld around has headphones on, has the person really called out at all? Earbud embodiment presents a very different future sound experience in the lifeworld than the past experience has been. This future is very personal and individualized with a playlist that fits every lifeworld scenario. The sounds of the world make a difference. They can create community. Murleau-Ponty describes sounds as objective, atmospheric, and reverberating and then permeating. But these degrees of sound can also be applied to the digital variety heard through earbuds. What is the distinction?

> Similarly, there is an objective sound which reverberates outside me in the instrument, an atmospheric sound which is between the object and my body, a

sound which reverberates within me "as if I had become the flute" . . . and finally the last stage in which the acoustic element disappears and becomes the highly precise experience of a change permeating my whole body. (Merleau-Ponty 2000, 227)

The atmospheric soundbed of the world is different from an earbudded soundscape. Both might be chosen and embodied, reverberating and permeating, but the earbudded one is personally selected so it is more exclusive of any random sound. Atmospheric sound encompasses all of the lifeworld sounds available at any given moment, with surprises, both good and bad, interrupting the senses at every turn. But earbudded sound is almost always chosen and selected and individualized. Surprises are rare because they are plugged out of the ear canal. These two listening experiences are very different. Once plugged in, we eschew atmospheric for individualized selection. The nature of community is changed by who listens to the world soundings and who does not. Earbuds have affect.

Earbud embodiment is one of many technologies that change human-technology connection. The opportunity to explore the technological artifact called the earbuds further illustrates the technological texture in our world as we dwell in digital sound. Chapters 7 and 8 complete the focus on digital audio with an analysis of the portable sound experience of the iPod, which is facilitated by earbuds, and an exploration of the technology and culture of a genre of music called dubstep.

NOTE

1. A portion of this chapter was presented as an academic paper at the 2009 Social Study of Science (4/S) International Conference in Arlington, Virginia.

Chapter Seven

Case: Portable Sound

Humans have used tools for as along as humans have existed. Tool use is a well-studied lived experience in many disciplines like anthropology, archeology, animal behavior and philosophy. Portable sound devices, like earbuds, are a tool. Portable digital sound, one of the reason we use these devices, is also a fundamental part of the human–technology connection. While chapter 6 specifically studied earbud embodiment, this chapter will delve deeper into a variety of facets of portable digital sound.[1] In the public sphere today, earbuds are attached to a device that facilitates music playing or an application that streams music. Sometimes the device is a mobile phone, but in many cases, the experience is facilitated through an iPod device, especially for those too young to have a phone. Older iPods played playlists made by the user, but the iPod Touch facilitates streaming audio apps as well. Similar digital sound devices flooded the music listening market but all of them are now on a steady decline as the mobile-phone market envelopes portable sound into their products by adapting the same contained music files and associated applications. In addition, although they are still less popular, users are turning to Internet radio and podcasts. To be sure, "[M]edia corporations rely on shared subjectivity of musical experience not just to sell as much of the same music to as many as possible but also to involve us emotionally in the films and games they produce, to help market the products they want us to buy, and even to sell us as a target group, defined by commonality of musical taste, to advertisers" (Tagg 2015, 2). Thinking about this human–technology connection of music wearing devices and all kinds of portable sound, especially music, through the lens of postphenomenology, invites new layers of understanding about digital media.

David Michael Levin, in *The Body's Recollection of Being: Phenomenological Psychology and the Deconstruction of Nihilism,* explores a deeper understanding of the human body "as a phenomenon of a *field* of being: an opening and a clearing" (Levin 1985, 65). While the previous chapter specifically focused on earbud embodiment and the specific technology of the earbuds, this case focuses more closely on the field of portable listening and music listening, and specifically, music-wearing, within an embodied field of being. The human who is using portable sound exhibits an opening and a clearing for communication, but what exactly is the portable device communicating? Michael Bull, among others, has done wide-ranging studies on music wearing (2000), iPod culture in urban environments (2004), mobile listening (2006) and iPod culture (2008). His ideas give us much to consider in terms of digital listening. At my university almost every student is connected as they walk across campus, wait for classes to start or sit down to eat lunch. In the library, in the gym and student union, at a sporting event, on the sidewalk, and waiting for the bus, ears are plugged in to a specifically negotiated soundscape that I am not experiencing. Yes, users young and old load music onto mobile phones, tablets, and laptops, and yes, many have stopped using the music player in favor of the "one device" experience, but there still seems to be a favoritism for the Apple music playing device. For many young people and preteens who do not yet own a cell phone, an iPod Touch, tablet or other digital device is the initial foray into portable sound. This case explores historical and cultural variations of sound and then moves on to micro-perceptual (embodied) and macro-perceptual (sociocultural) understandings of portable digital sound.

Let me start with three assumptions. The first assumption comes from the philosopher Don Ihde and a host of others who stress the non-neutrality of technology. As Ihde reminds us, in *Bodies in Technology* (2002), humans and technologies are both changed in the interactions, but humans have intentionality. He says, "My middle ground claim is that there is, indeed, a limited set of senses by which the nonhumans are actants, at least in the ways *in which interactions with them, humans and situations are transformed and translated* (94)." So, I will put forth the premise that an object, like music-wearing technology, modifies the human listening to it and the human experience in the world is changed so the experience is translated into a different kind of being-in-the-world. The second assumption is the power of the aural. Ihde notes that "cognitive functions of listening are more complicated than seeing, and sound plays a stronger lifeworld role than vision in some cultures" (2002, 41). Levin, in *The Listening Self: Personal Growth, Social Change and the Closure of Metaphysics* (1989), suggests that "Hearing is intimate, participatory, communicative; we are always affected by what we are given to hear. Vision, by contrast, is endistancing, detached, spatially separate from what gives itself to be seen" (32). And the third assumption is that the lived experience for a portable sound user is grounded in perception, and perception is rooted in the body. As Merleau-Ponty explores in *Phenom-

enology of Perception (2000), "I regard my body, which is my point of view upon the world, as one of the objects of that world" (70). Assuming that portable digital sound is not a neutral technology, that portable sound can be a powerful experience, and that listening is grounded in perception, allows for a clear jumping off point to consider the lived experience of dwelling in portable digital sound.

I wear my portable sound device on my body. My attention to my environment, my being-in-the-world, is the awareness I experience through my body. I might existentially feel that my body is absent in the world, that time has flown, that I am in a non-bodied place, but perceptually, my life is about a connectedness between field and ground in the world through my body. But as noted in chapter 6, if I have earbuds in, the connectedness to the field (the lifeworld) is different than if others can hear my portable sound or if I can also hear the lifeworld sounds as I listen to my portable sound. Additionally, when it comes to music listening, my body is primed to respond to the content. Moods and heartbeats reach to the tunes of the chosen playlist. Human–technology connection is situated, then, in the conception that our experience in the world is changed and our being-in-the-world is different when we listen to portable sound. The aural is a strong communicative force in our world, and listening occurs within our bodies, our "view" point of the world. But how can this affect be named?

Postphenomenological variations explain the historical trajectory of portable music players. The 1954 Regency TR-1 Transistor Radio was one of the first portable audio devices so portable listening has been around for some time. Michael Bull, in his ethnographic work *Sounding Out The City: Personal Stereo and the Management of Everyday Life*, observes that personal stereo users utilize personal audio devices, so they do not have to interact with others or the environment, becoming "nonreciprocal," "one step removed" from their surroundings, and "withdraw into themselves" to make their music or other information "all encompassing" (2000, 25). A variety of mechanisms, from Billboard to iTunes to Internet radio, negotiate the statistics of the music content that is downloaded or tuned in to. Music can also be purchased on CD and imported into the player, but the majority of portable music is downloaded.

In the last few years there has been increased conversation about a kind of music sharing called iPod jacking. The action does not involve specifically sharing earbuds, but unplugging the earbuds from the jack of the iPod and sharing the song with someone else, which can last a few seconds or a few minutes. One of my favorite photos shows two students sharing an iPod, each with one earbud in their ear. This is a more personal and bodily close sharing. But it does not happen often. From Internet discussions on the iPodlounge to articles in popular blogs and magazines, it is clear that an abundance of technology is being built and marketed for personal music listening. Hence the "I" in iPod and the variety of "I" inspired technologies. Embodiment and culture are changed by this portable digital soundscape.

1 CULTURE

An iPod can also be explored through signifier and signified. The icons and display on an iPod, for instance, share a language of many years of audio technology and audio culture embedded throughout the history of electronic and digital media. The history of that industry has made an indelible mark on the present. Turning the device on and off at the edges, negotiating menus, turning the click wheel, selecting a song, and playing with the volume all conjure up a variety of historical sensitivities about audio, from early radio to computer automation, through the metamorphosis of audio technology from cart machine to digital playlist. For many years industry experts proclaimed that radio, as an industry, was dead. The socio-cultural markers, overlaps from a prior culture, if not the actual technology, have been reborn and carry on an entire tradition. Internet stations use the same language and programming structures as legacy radio. Of course eventually a link to terrestrial radio might be gone for good, but for now, this link is a fruitful connection to the culture of the radio industry and of pop culture. My first digital portable sound device experience was not what I'd call intuitive, but I learned it quickly. For me, the semiotic similarity to non-digital technologies provided a pathway for understanding. Semiotic icons bear resemblance to the "what" they represent. Brand identity has also smoothed the transition. The Apple iPod identity represents a file-sharing computer database, an audio technology through icons, a cultural and brand. Apple iPhones continue much of the same flow and now the Android operating system for the cell phone also has a similar workflow for music. The way I read my technology and the socio-cultural experience of music are changed by the radio history.

Further overlap can be seen through the variety of ways digital radio has found a place in the world. Downloaded music has changed the recording business and allowed music artists to share their own music without going through the long-held traditions of the record and recording industry. Music sites like SoundCloud facilitate sharing portions of sounds to create new ones. Music applications like SoundHound allow users to identify songs to facilitate their download and purchase. Digital downloads have given local artists national exposure. And the live concert experience continues to have a strong appeal for many.

Vivian Sobchack suggests that the perceptive body is an iconic sign. "Perception *is*, therefore, intentionality-in-existence as it *presents* itself to the world and in it as the concrete, existential manifestation of intentionality. My perceptive body *is* my intentionality in the world, is in existence and as representation identical to it" (1992, 74). We have intentionally taken up a position as a portable sound device wearer and the gesture perceives the world through this kind of lifeworld being-with. The intentionality-in-existence is a representation of one intertwined with technology and promoting the Self in decided one-ness.

The process of music wearing is macroperceptually part of contemporary culture. Personal ownership, individual listening, non-verbal gesture and technological connectedness usually mark the sign of the music wearer. The devices and the culture are music wearing ready. Music-wearing gear is an acceptable cultural norm. People understand the process of downloading files and most devices facilitate the downloading process. Most people do not expect to hear someone else's iPod on the subway or in the gym. Society has etiquette around sound volume. As Sabchack explains, conventional symbols are marked by "perceptive and expressive persons coming together in cultural agreement whereby the relation between the signifier and its signified is arbitrarily made to stand as a sign" (1992, 75). The materiality of the iPod, entwined in the person, represents the body and the person together as one entity, and the music wearing and music-listening as "mine."

When I plug into the iPod jack and pop the earbuds into my ears like an accessory or piece of jewelry, I produce a gesture called "plugged into and listening to my personally chosen soundscape." Music wearing signifies time and place connections; a physical gesture toward being-in-the-world as a listener-of-personally-selected-audio-that-only-I-can-hear. As Sabchack notes, an indexical sign is in relationship with its signified. The iPod is close to my body, fits into my pocket or backpack, clips to my shirt or is otherwise worn on my body. It is connected to me and I experience it through my lived body. It is wrapped around me and I am plugged into it as it plays my music or my audio file. I intentionally put on my iPod and go out into the world. When I am home, I may use the speakers while cooking dinner or working out. My music wearing body points, says Sabchack, "to the world's presentness to and for consciousness, and for consciousness's presence to and for the world. It is the means of connection whereby I can say and know I *have* a world" (1992, 75). It is interesting to think that consciousness is concerned with being awake and aware of what is going on around me and occurring within the field of my music-wearing self that is potentially "un" conscious of the natural world sounds.

I recently came across this orienting quote in musicologist Phillip Tagg's book *Music's Meanings: A Modern Musicology for Non Musos* (2015). One of Tagg's big questions intrigues me and guides this case. Tagg writes, "Why and how does who communicate what to whom and with what effect (41)?" This phrasing, based on Harold Lasswell's 1948 communication model, explores the idea that music is a non-verbal system for mediating ideas, one that is difficult to articulate effect (40). Breaking down the communicative process helps to loosen the technological texture of portable digital sound. The next section is guided by the seven different parts of this quote, to more fully open reflection to further consider the postphenomenological framework of micro- and macro-perception.

> **Why** and how does who communicate what to whom and with what effect?
> Why and **how** does who communicating what to whom and with what effect?
> Why and how does **who** communicating what to whom and with what effect?
> Why and how does who **communicating** what to whom and with what effect?
> Why and how does who communicating **what** to whom and with what effect?
> Why and how does who communicating what to **whom** and with what effect?
> Why and how does who communicating what to whom and **with what effect**?

The Why

Why did I want to get an mp3 player and then an iPod? Why was it important? Why did I want to spend the money? And why do I still use these devices when a mobile phone can play music, too? I'd suggest that I like a device specifically to collect my music, I like communicating that I want to be by myself, that my music is more interesting than anything else, that I prefer to be shut off from accidental noises that I have not invited in, and that I want planned sounds in my world. I can control my environment in a certain way, with "noise" that I have paid for. Why am I communicating this? Because I want to manipulate the sound around me, because my choices are pleasant to me, because I want to make the best use of my time by programming my day with my choices to get more done, to manage my stress, to elevate my mood, to narrow the margin of having to deal with anything unpleasant. I want to personalize my environment to the optimal level.

The How

How am I communicating? My nonverbal cues suggest that I do not want to interact. Unless, of course, I'm inviting—through gesture—someone else to jack into my iPod, which is often considered jarring and encroaching if it happens unexpectedly. The iPod technology allows me to select a specific worldly stance because, again, I am on the move, not impeded with large audio equipment, not sharing music with others, but plugging up my ears so I do not hear anything from others. IPod users have shared that they will turn off their audio but keep their earbuds in their ears so they do not have to deal with any kind of verbal communication and often promote non-verbal language that further closes off communication beyond the plugging of the ears. Plugged ears say, "Don't bother me, I have decided not to listen to you because my sounds are more important and the world's soundscape is too noisy." I turn the iPod on to intentionally manipulate the soundscape of my life. If someone does want to get the attention of the iPod user, they need to use broad gestures, touch them or speak loudly. These are all more invasive and less emphatic forms of communication.

The Who

The *who* of the question is the music wearer, the iPod user. The *who* are users the world over who own iPods. And many people do not just own one device, but have three or four different devices that overlap. The *who* are individuals who intentionally choose to intertwine with a technology and plug it in to their body. The phenomenological intention of the world, taken in and apprehended along with the sounds from the iPod, makes up the music wearer's world. And there are many cases where music wearers do not want to be accessible through a smartphone, but do wish to listen to music.

As Merleau-Ponty explores, in *Phenomenology of Perception* (2000), one's "phenomenological world is not pure being, but the sense which is revealed where the paths of my various experiences intersect, and also where my own and other people's intersect and engage each other like gears" (xx). How might these intersections occur for the portable sound wearer? How is one's "rational experience in the world laid out" when "all cognitions are sustained by a 'ground' of postulates and finally by our communication with the world as primary embodiment of rationality" (xxi). We are choosing, rationally, to plug our ears and choose specific sounds, turning away from lifeworld sounds occurring in our phenomenal field of experience.

Music wearing is a habit. Merleau-Ponty (2000) suggests, "habit in general enables us to understand the general synthesis of one's own body" (152). How might the *who*—the music wearer—perceive the world and communicate? For the *who*, the iPod is no longer an instrument, a music player to wear, but an "instrument with *which* he perceives . . . It is a body auxillary, an extension of the bodily synthesis" (152). And there is an effect to the synthesis. Merleau-Ponty's insight here helps clarify the effect. He writes, "It conceals the organic relationship between subject and world, the active transcendence of consciousness, the momentum which carries it into a thing and into a world by means of its organs and instruments" (152–153). The new apprehension, through the instrument, the iPod, enriches and recasts the body image. Experiencing the lifeworld comes in and through the digitally translated math problem turned music mediated by the technology called the iPod, plunged in to the world.

We can take the music-wearing experience as a sensory experience, which then becomes a unity of the senses and a unity of the world with the music-wearer's world. There is no distinction between truths, only the world actually how it is . . . the fact that we are in the world . . . and "every sensation is spatial" (221), that we are bodies in the world, and that we are coexisting in a place together. The sensory experience is a form of existence. The iPod taps the aural sense. Because I have tapped into the sense, I, as Merleau-Ponty explores, "can never see or touch without my consciousness becoming thereby in some measure saturated, and losing something of its availability" (221). I am saturated by the soundscape. How is it then, with the sense of hearing? He adds, "It is neither contradictory nor impossible that each sense should constitute a small world within the larger one, and it is even in virtue of its particularity that it is necessary to the whole and opens upon the whole" (222).

The Communication

Music wearing solidifies the priority of the personal and the private. People don't often share while music wearing. While they don't share the technology, they do share worldly space and ideas about what to download and what to buy that might accompany the brand. IPod wearers might share a bench on the bus or subway or the sidewalk on a city street. They might share a side of the room in a gym. These are "communal" and singular experiences. People relate in different ways when they are co-mingling their small world sounds among the larger world of sounds. The bottom line for the music wearer is that the user's ears are covered or plugged by earbuds. The digitally created music takes the auditory priority in the lifeworld. When the earbuds are connected, the iPod user is showing that he or she has intentionally given the chosen sounds coming form the iPod the priority. Others in the world are the background automatically, until they force themselves into the foreground or the iPod user chooses to shift intentionality.

The technological instrument becomes the mediator of the world (Merleau-Ponty, 2000). The world sounds become mediated through the music coming from the earbuds. Anything that gets through receives attention, and anything that does not come through is not attended to, based on the priority set by the wearer. Just as an instrumentalist habitually plays the instrument "without thinking," with the music becoming its own thing in the world, the iPod audio becomes that thing that the music wearer mediates the world through. Merleau-Ponty (2000) writes, "Sometimes, finally, the meaning aimed at cannot be achieved by the body's natural means; it must then build itself an instrument, and it projects thereby around itself a cultural world" (146). Might the iPod be a cultural instrument too? And how might the content of the music become part of the way we mediate and negotiate the world?

The What

What is communicated between the sender/music wearer and the receiver/world? Etymologically, *communication* can mean an exchange of information between individuals, through a common system of signs and behaviors or through a sense of understanding and sympathy. In our highly technologized world, the music wearer's stance in the world can be interpreted in many ways. People have many thoughts about those who wear an iPod. The *what* is dependant on a variety of interpretations. The iPod-wearing stance could be considered escaping the larger environment by creating a sub-environment, or finally arriving as part of the crowd. Someone could demonstrate this by putting earbuds in the ears. Another person might spatially separate themselves by going off into a corner away from others. Someone else might choose to absorb the sounds around them and watch others.

The Whom

Others in the lifeworld are the *whom* in this semiotic process. A person in one's lifeworld can be friend or stranger, acquaintance or lover. What boundaries does a person make when using a portable sound device with earbuds? How important is it to be present to sounds in the lifeworld? Just by plugging my ears with my earbuds and plugging in to my device, I am plugging up my option to participate with aural sounds in the lifeworld. Have you ever finished an argument by plugging in to your portable sound? What does that say to the person you were communicating with?

With What Effect?

What might be the effect of portable sound? We might not know this for some time. Years of research has focused on the experience of music listening, but the human–technology connection of music wearing is different and less studied. Technology concerns about hearing loss because sound is plugged into one's ears and issues when pacemakers and other health devices are used, have been discussed in mainstream medical journals. One less studied potential aural effect is through digital drugs focused on sound. I–Dosing, listening to sound focused on binaural beats that change brain wave patterns and induce altered states, is slowly gaining momentum and popularity (Lavallee, Koren and Persinger, 2011). While widespread research on digital drugs is not yet available, studies share that binaural beats can manipulate cognitive processes and mood states and create an auditory illusion beyond that of general music listening (Chaieb et al, 2015; Reedijk, Bolders and Hommel, 2013). The website I-Doser sells music download *doses* aimed at producing simulated mood experiences through music simulations called brainwave digital doses. Each device, and each experience adds residue to the human–technology connection.

MICRO/MACRO BUBBLE

Glen Mazis explores this nexus in *Humans, Animals, Machines: Blurring Boundaries* (2008). He writes, "An invocation to openness of experiencing often invites the question, 'whose experience.' Experiences seem to vary so widely given personal, historical, cultural, ideological, and other differences" (13). He discusses that perception is our way into the world and explains, "Our perception and overlapping feelings, emotions, memories, imaginative echoes, and so on are not 'our accomplishments but co-accomplishments' with all those beings to which we relate. Perception is a gathering together of all of those levels of meaning" (15). How might portable sound change the space of our perception, our surround? How might we experience different auditory dimensions of the lifeworld as a human–technology connection? I wonder, as my foreground and background perceivably shifts, if I am shutting out, bringing in or just apprehending the world differently. What might be the residue, the remainder? If my plugged-in sound gives me comfort, am I a happier or "better" citizen in the world? If my music helps me work through the multiplicity of choice in the world, I am more equipped to negotiate the lifeworld. If I cannot hear lifeworld sounds, am I less of a citizen? Mazis says, "In order to have real communication among realms, there has to be seen both overlaps and boundary" (27). Music wearers create boundaries by wearing earbuds linked to technologies but do they create overlap as well? Postphenomenology is a helpful framework here to separate and explore the nonneutrality of this human–technology connection of portable sound devices by focusing on the micro (embodied) and macro (socio-cultural) perceptual experience.

My portable sound device is part of the lived world, and the sensory world and my soundscape is most centrally composed of music. I wear my music and this field of being also affects my presence and perception of the world. Merleau-Ponty calls this space the natural or common world, and explains that both affect our total experience in the world.

> Music is not in visible space, but it besieges, undermines and displaces that space . . . The two spaces (lived world and sensory world), are distinguishable only against the background of a common world, and can compete with each other only because they both lay claim to total being. They are united in the very instant in which they clash. (2000, 225)

He explains that concentrating on a specific sense becomes "his world for a moment" (225). He suggests that all senses make up our natural or common world, and by focusing on one sense, like the sense of hearing music through a device, "one assumes a highly particularized attitude" (225). What might this attitude mean in the lifeworld? How might boundaries and overlaps occur? A soundscape based on my personally programmed playlist of downloaded music creates a particularized attitude that spills over to all of my other senses. Merleau-Ponty further explores this union of the sense by clarifying that

> synaesthetic perception is the rule, and we are unaware of it because scientific knowledge shifts its centre of gravity of experience, so that we have unlearned how to see, hear, and generally speaking, feel, in order to deduce, from our bodily organization and the world as the physicist conceives it, what we are to see, hear and feel . . . The senses intercommunicate by opening on to the structure of things. (2000, 229).

This micro-perceptual lens of feeling and organizing the world through our body, coupled with the macro-percpetual structure of the world, bump and rub and weave together to become entangled. Gunn and Hall, in their 2008 study, "Stick It In Your Ear: The Psychodynamics of iPod Enjoyment," describe the iPod experience as a "sonorous envelope," a naming of losing oneself in the music. My music taste, then, is co-mingled with my perception of lifeworld experiences. I am within the bubble but also within the field of the lifeworld as I move in the bubble. This is macro-perception.

Micro-perceptually, when I am "wearing" my music, I think my lived body communicates many things: "I am busy listening to something," "Don't interrupt me, I'm exercising," "I don't expect to know you so I am going to create my self-selected world sounds in this unfamiliar place," "I'm bored with the lived world's sounds so I'm going to listen to my chosen audio field," or "I feel safe in this environment because I have some control over a portion of my personal experience." Earbuds say personal, singular and individual. It wasn't called a Walk "men"—but a Walk "man." First the Sony Walkman and then the Sony Discman, held my personal music tastes until the iPod came along, with its seemingly limitless choices for listening and an opportunity to get away from the "man." The portable sound device has changed both embodiment and culture by its invention. The trajectory and the wearing all explore the experience of portable sound devices and their certain effect in the lifeworld in micro- and macro-perceptual ways.

Chapter 8, where I consider the dubstep genre, completes case studies that consider digital sound.

NOTE

1. An initial study of this topic was shared in a paper at the National Communication Association in San Diego, California, on 11/23/08.

Chapter Eight

Case: Dubstep

Each case in this book is an opportunity to probe for what is genuinely discoverable but not often seen. More than mere analysis, a postphenomenological rendering of a human–technology connection opens the door to unseen possibilities "experienced and verified through the procedures that are, in fact, the stuff of experimental phenomenology" (Ihde, 2012 13). Chapter 5 reflected the experience of digital sound, and chapters 6 and 7 explored embodiment through wearable sound technologies like earbuds and portable sound. This final case in digital sound builds on the three previous cases by looking specifically at the performance and making of digital sound using digital tools, with an aim of probing the role of music technology in personal, social and cultural life. The human–technology connection of the dubstep nation means language creation, performance, peak experience and collaboration.

The human body is a perceiving participant in the human–technology connection. The dubstep genre of club music illustrates technological texture through its highly processed style of electronic dance music (EDM) played in nightclubs or concert venues and warehouses, and largely created by and for DJs who remix electronic music in a variety of ways that evoke a live performance feel. Two overarching ideas are illustrated through dubstep. First, any technology can be used in a multiplicity of ways, limited only by culture and human ingenuity (Ihde 1995, 37). And second, technology is a cultural instrument. For example, tools used for basic sound mixing or tuning, like Auto-Tune, can become a technology for music mixing to the degree that a new genre of music is created. It is not a surprise that both dubstep and Auto-Tune became official entries into Merriam-Webster's Dictionary 2014 edition. Auto-Tune is an audio processor that is connected to a computer. It is used to alter and measure pitch in vocal and instrumental performance. At first Auto-Tune was (and still is) used for pitch correction. Soon the auto processor was used to treat or process voices in a variety of ways, like applying distortion to the mix. Auto-Tune and other technology play a role in the creation of dubstep. After this case we move on to several other illustrations that continue to explore the non-neutrality of the human–technology connection.

Wired for Sound

This current genre of dubstep is often traced back to EDM, the Electronic Dance Music genre that has been around since the late 1970s. The EDM collective includes genres like Bass, Big Room, Deep House, drum&bass, Chill Out, Future House and Hard Style, and are predicated on the fusion between technology like computers and synthesizers and DJ in a human–technology connection that is hard to beat. Each of these diverse dance tribes, as they are sometimes called, shows up on large stages around the world, but also can be located deep in the underground of music and DJ culture. This music, along with its performance component, has been well explored in Van Veen and Attias (2011), Pfadenhauer (2009) and Dekker (2003). Electronic Music Culture (EMC) has a large and growing population of fluid and engaged fans whose participation goes far beyond the term of listener. One of the more recent movements, or subgenres, of EDM is called dubstep. This community is enthusiastic and passionate, and their beats and hooks are infectious. But the history of this genre is contentious and what constitutes "dubstep" is under continual scrutiny on fan sites and popular media. No one is even sure when and where dubstep started. Many agree, however, that dubstep is the sound of a generation. Strands of DJ performance, a new language for digital tools and music composition, and DJ and fan collaboration will be considered in this case. Although some music professional say that Techno, Electro and Deep House genres are taking over as more popular forms of EDM, the quick rise of dubstep is still worth studying. Musicologist Phillip Tagg (2015) explains "The sounds . . . of electronic dance music posed a whole range of questions about a major shift in patterns of musical structuration, questions that cannot be answered by merely pointing to development in audio technology (MIDI, synthesizers, sampling, etc.) and to young people's use of that technology (444). This reflection takes up Tagg's questions and aims to uncover what is often not seen or heard, from the "dubstep nation."[1]

Why Study Dubstep?

My turning to the phenomenon of dubstep is partly personal curiosity. I hear dubstep around me in clubs, in movie and video game soundtracks and in commercials. The sound is different from prior forms of EDM. Music magazines and websites talk about the "godfather" of dubstep and the "early adopters" of the genre, and mainstream music has only recently embraced the genre. Dubstep does not regularly top the music "hit" charts but has been a success at clubs and on various sharing websites for a decade. EDM and EMC rating sites like *beatport* share the top mixes of the genre and have reported both the rise and fall of the genre regarding sales. The genre is neither underground nor established. It has made its own mark and own independent market for music while avoiding mainstream media in favor of the becoming the kingpin of the club scene.

My DJ friends perform dubstep mixes at clubs and parties, and I have been considering how this kind of music is different from previous forms of electronic music. And I must confess, I have always had an interest in sound. I've been a DJ at several radio stations and taught digital audio production classes at my university. My work in digital media allows me to "play with" a variety of digital audio tools, and I had the opportunity to learn to use a MIDI in a graduate school class to create a soundbed for some visual work I created. I enjoyed the American East Coast rave scene in the early 1990s but stopped clubbing to get my doctorate and have three children. And, truth be told, I married the sound guy in my brother's band. So I turn to this case with sincere interest, curiosity and a bit of understanding. This case only scratches the surface of the complexity of EDM and EMC culture, but uses postphenomenology to clarify interesting trajectories and variations of this human–technology connection.

This case requires a few caveats and reminders before further study. First, it is important to remember that technologies are non-neutral. "They are transformational in that they change the quality, field, and possibility range of human experience, thus they are non neutral" (Ihde 1993, 33). Second, "Technologies must be understood *phenomenologically* i.e., as belonging in different ways to our experience and use of technologies, as a human–technology relation, rather than abstractly conceiving of them as mere objects" (34). And third, the technological transfer is multidimensional in that it will "involve basic cultural and existential interchange" (34). One additional layer deserves mentioning: an understanding of the intentionality or "trajectory" of technologically embedded experiences. The myriad of technology used in the creation of dubstep mediates the relationship between DJs and their fans and listeners, and plays an active role in its creation and context.

> When technologies mediate the intentional relation between humans and world, this always means from a postphenomenological perspective that they codetermine how subjectivity and objectivity are constituted. Their intentionalities, in Ihde's sense, consist of the fact that they co-shape the contact be-

tween human beings and their world: they determine how human beings can be present in the worlds, and the world to them. (Verbeek 2005, 6)

The mediation of the subjectivity and the objectivity is a key element in this case. How DJs and their technology co-determine and co-shape dubstep is an important consideration in the DJ (human) turntables, computer, sharing websites (technology) fans, audience, sharing websites (world) relation.

HUMAN–TECHNOLOGY WORLD

Dubstep is an example of technological co-shaping between humans, technology and the lifeworld in one giant feedback loop. DJs create mixes and beats using sound software programs and other technology like turntables and Auto-Tune, that they share online in music sharing websites or with live audiences. Sometimes mixes occur live and are choreographed in some sense by the enthusiastic swell of the crowd based on DJ mixing on the fly. Then, these mixes are often loaded into software and onto sound websites. This process of DJ-software-live music venue or DJ-software-sound website is co-shaped to the point where it is hard to distinguish where the sounds started and who created them. The aim of this case is to study the co-constitution to search for a deeper multistable structure within the possibilities (Ihde, 1993).

The dubstep environment is one of listening, mixing, remixing, performing, live mixing, listening, moving, remixing the live mix, dancing and experiencing. Has dubstep become a new cultural context for the technology used for this genre? Has one of these "post" genres, called brostep, a more Americanized version, seen technology transfer based on a new set of material artifacts available to DJs, the largest group of dubstep music makers? And regarding the technology, how do the technology and the dubstep artists co-constitute the experience within the lifeworld. And what of the newer iterations called brostep, post-dubstep and purple dubstep? For Ihde, (1993) technology is a cultural instrument only limited by the cultural imaginings of the individual. Indeed, the material music culture of dubstep, through human–technology-world interaction, has imagined and created new cultural dimensions for electronic dance culture.

In the mid to late 1990s, a group of DJs started experimenting with new computer based sounds and inputs and outputs in garages and homemade studios in South London. This movement was started by British sound-system culture and cultivated through a bass centric focus in London clubs. The digital sounds eventually trickled into warehouse dances and clubs to the point where other digital music creators, though less sophisticated in the ways of building these digital mixes, started experimenting too.

After dubstep found a homeland in the UK, it spread throughout the world and created subgenres that shadow but do not reify the original goal, while adding new and different dimensions to the sounds. The combination of affordable equipment components of varying levels of complexity, website platforms to post dubstep for sharing, a ready crowd to critique the music on sites like SoundCloud, beatport and subnav and sharing on social media and Youtube once a DJ's set was ready for the big reveal, made dubstep a popular pastime for both creators and fans.

Like a band, Dubstep is created from a set of songs, beats, and previously recorded "mixes" built from a loose playlist based on the interest, requests, and swell of the crowd as an organic feedback mechanism. But crowd swells at live concerts have been around for a long time. Film clips of the Beetles, Rolling Stones, Elvis and U2 bear witness to the power of crowd enthusiasm and interaction. What makes dubstep different is the continued doubling over and remixing of the raw material into a derivative work. This synchronized segue, an individualized mix, is often connected with the 1990s rave dance trend and is the cousin or child, depending on who you ask, of disco, house, techno, hip hop and Jamaican DJ cues. Popular dubstep artist Skrillex collaborated on several projects with artists who largely work in other genres with many different types of people on dance music that ranges from ravers to hip-hop (Herrera, 2011). The combination of already-created music and new sounds, remixed and recreated in different layers live by a DJ, moves the production and mastering phase into perpetual loops as the performance of the material becomes part of the digital mix. That digital mix is again tweaked, uploaded, shared, potentially downloaded by others to use in a mix and then possibly played live again for another loop of digital interaction.

At its essence, dubstep is a music mix combining the use of sequencing platform software programs like Logic Studio or Dub Turbo, with an Auto-Tune plug in, and speakers with large sub-woofers to hear the bass and drums, a beatbox with a 140 beat-per-minute timing, and the chorus of a song or two as a hook to keep people familiar with the song as they react to it within the club environment or venue.

Part of the allure is wobble bass or web, used frequently in dubstep, as an extended sound that takes on a rhythmic tone. A low pass filter with a low frequency oscillator assigned to the cut off proliferates the distinct wobble of the bass, which combines with the drum. The sound is facilitated by the use of large speakers technologically able to transform the low sound into amplification for the audience. Early dubstep, especially, is rarely played through earbuds, because much of the low wobble of the sub-bass is lost because the tiny speakers are technologically unable to transfer the low tones of dubstep that can be felt by listeners in a very bodily way. At its best, dubstep is a kind of music created to be heard in environments that have the speakers that are able to capture and amplify that low sub bass range.

MIX, MASTERED

DJ, technology, sounds, and the audience co-create dubstep. A variety of technologies are used in the digital creation, but the final mix is output on social media sites and uploaded onto dubstep sites for regeneration into another DJ's mix. These mixes are played live at clubs and in the studio. The mixes become fluid interweavings of what is created, what can be used from others and what can be manipulated within one's own dubstep workstation. Dubstep uses many of the same digital software tools and vocabulary as EDM, so the proliferation of what Ihde calls *"familiarity within a known praxical gestalt"* (1995, 35) occurs within the dubstep movement to create subgenres. The mixing and mastering of dubstep occurred along the same known praxis of other EDM, but new sounds were created in the process. The consistently recognizable and understandable interface of the variety of music-mixing tools available, like propellerhead, Deckadance, Live and Push, digital audio workstations like fl studio, and many free downloadable sounds, paired with easy creation of mp3s uploaded to sharing sites, fosters the genre and EDM in general. Synthesizers, drum machines or beatboxes, sequences and synthesizers, along with vocabulary like sampling, scratching and tracking, drops and hooks, have become universal to DJs, fans and electronic music brands as well as the musicologists, historians, and cultural anthropologists who study digital music.

The compound hearing experience of dubstep fragments, feedback loops and electronically generated sounds, pulled together in space and time, constructed on the fly, from components mixed over time and new elements, means a song turns into a set and turns into a twenty-minute "mix" or entire "set" of music that occurs before a "break." One thing is sure, a population of opinionated dubstep-loving people is inhabiting the planet and have probably already begun to disagree with the ways in which I have explained dubstep or categorized its place in EDM history. And that is par for the course for the genre and its lively fan base.

THEORY, CONSIDERED

Variational theory, which uncovers the variety of multistabilities of material technological artifacts, is one of the key concepts of this reflection. Ihde employed the concept of multistability as a way to explain technological mediation. This insight paved the way for postphenomenological research as an approach to understand the transformation of the perception of an object. The idea is to uncover variations in the mediating role of dubstep for the audience and the DJ to reveal the multiple ways that the music technology facilitates, fosters and creates dubstep, in sync with the embodied DJ and audience, to create a "new" music mix, that is posted, downloaded and again doubled over on itself as a new performance experience. The mixes that are hits are uploaded, voted on, and downloaded and used again, in a way that erases much of the ownership of anything original, for the newness of the most recent mix.

The process emphasizes essential structures or trajectories (Ihde, 2009) to determine what is variant and altered and what is invariant or unaltered. It is through beginning insights of a technology's use that we can have, notes Ihde, "insight into the very structure of multistability, an insight that then can guide our subsequent awareness, such that we might well expect both more possibilities and . . . multiplicity of profiles" (Ihde 2006, 201). Thinking through the human–technology connection can help develop a way of constituting ourselves in this mediated world, to reveal a deeper and more rigorous analysis of the illusion of the digitally altered body engaged in the wub, wub, wub of bass, with drops and pops of the hook, loops, amidst the strain of a pop song.

Postphenomenological variation and inversion can be uncovered by thinking about the micro-perception and macro-perception of the music genre, with its focus on bass, griminess, dirty sounds and hooks. A micro-perceptual lens highlights the sensory experience of feeling the beat of the bass, recognizing a familiar hook, moving in sync with the crowd and having fun in a culturally attuned ritual of dance and movement. The gaze is intent on taking the experience in and viewing the lights, which often mirror the beat and audience movement. Macro-perception, conversely, discovers that this specific perception, this sound, this view, this look, this body, can be completely appropriate and attainable. Contextualized, there is a cultural shift of accepted experience. What is revealed in the essences or variational elements of *new language, performance, peak experience* and *collaboration* through the experiences of those who dubstep?

Source materials for this reflection have been gathered from magazines, Internet interviews, websites and a transcribed interview to consider the dubstep experience. Experiences of EDM and dubstep from a performer and producer, California native Skrillex, and from a South Central Pennsylvania local spinner, DJ Merc, provide a context for the dubstep experience. Together, multiple ways of relating and communicating in and through dubstep can be considered. After reviewing the history and technology of dubstep, we can see it as more than music to listen to. Dubstep is a new language, a performance, a peak experience and collaboration that clarifies human–technology connection in electronic music culture and explains dubstep as an instrument of culture.

New Language

Like any other cultural movement, dubstep has spawned a whole new vocabulary of terms. Reading fan comments on a website like SoundCloud is one way to explore the variety of vocabulary associated with dubstep. Many different blog posts reference the music side of the genre using terms like glitched sequences, Mini Mix, samples, rave synth, trance, trap slaps, beats, pops, bounce, riffs, jams, tempo, dirty, grungy, muddy, wob and wub. The technical terms (beat pad, track, leads, samplers, MIDI controller, 140 bpm, kick sample, punch, patch, sub, reverb and drop) and the Internet message board words like noob (someone new to dubstep mixing) and newbstep (dubstep mixed by a noob) also bring new language to the genre.

DJ Merc explains it this way:

> There is a really constantly loud predominate bassline, with pretty synthetic synthesizer noises, Auto-Tune beats throughout the song and obviously it goes to a tune, but there's no lyrics, it's all noise that make up a song and they can actually beat mix on the spot . . . throw in the hook.

Online sites like DSF Community and many of the brand name technology websites have places where DJs share their craft while mixing the language of brands, technology, music and message board culture.

Performance

Dubstep is, at its heart, a music performance. The DJs are digitally performing the dance of audio editing. They mix the beats, bytes and bits to the rhythm and cadence of interwoven sounds. They perform their understanding of the technology by negotiating the place of the software by pulling in different sound strands; laying them down on a digital track interface and then mixing them and compressing them to many types of files including AIFF, WAV, FLAC, ALAC, OGG, MP2, MP3, AAC, AMR, or WMA portable sound files and played on a website for comment, critique and listening or performed "live" for crowd. DJ producers knit old and new, using instruments like violins with dubstep machines like the Cubease 6 and Native Instruments. This gathering of sound acts like an open playing field of endless perceptual options of "experiments" using things "lying around." This is "*starting in the midst of* that which is already here, the already constituted. That is a beginning from the center" that is dubstep (Ihde 2007, 48). A dubstep DJ listens "to the *voiced* character of the sounds of the World" (47). Much of the music that is uploaded to music sharing sites tied to social media receives feedback and the statistics on listening and commenting provides instant feedback. Feedback is a valuable and social opportunity that allows DJs to reach out to the "live" listening community. The social element of

EDM and other digitally created music has philosophically been questioned and considered long before dubstep became a genre. Noted musician John Cage, in his essay, "The Future of Music":

> Technology has brought about the blurring of the distinctions between composers, performers, and listeners. Just as anyone feels himself capable of taking a photograph by means of a camera, so now and increasingly so in the future anyone, using recording and/or electronic means, feels and will increasingly feel himself capable of making a piece of music, combining in his one person the formerly distinct activities of composer, performer and listener (1979, 120).

Cage explains his concerns that the blending of the roles will take the social element from the experience, saying that, "to combine in one person these several activities is, in effect, to remove from music its social nature" (2009, 120). But this has not happened in dubstep. The sharing, the listening and the relooping of the mixes and the reloading and resharing based on audience interest, feeds off the social environment, and performed live or through digital sharing.

Peak Experience

Euphoric mental experience, excitement, and exhilaration are all part of peak experience. This carefree and aesthetic experience has been linked with dubstep and dance culture because the music is made to feel supernatural (Juslin and Sloboda, 2010). The field of embodied music cognition is working to link embodiment with peak experience. The thought is that the emotional experience intertwines with the fully digital production process to create an experience that fosters embodied experiences like being pushed upward, held in suspense, and dropped down, and the expectancy of experiencing these corporeal opportunities while listening to the music. A dubstep audience anticipates, comprehends, and apprehends the listening as they might enjoy other peak experiences like amusement rides or extreme activities (Solberg, 2014). The experience of "moving one's body to music releases dopamine, explaining why we may experience pleasure and well-being when perceiving aesthetic expressions. The club dance is characterized by being one amongst a crowd, as opposed to dancing one on one. It is both an individual and a social act at the same time, where the clubbers wish to dance" and become part of a crowd (78). The technology in the music production and performance processes, coupled with human participation in the process and expectancy of the music comingle in the human–technology connection. The additional experience of feeling part of something bigger than an individual experience also heightens the peak experience.

Collaboration

Collaboration is the name of the game in dubstep. There is collaboration in the making, the sharing, performing and experiencing. And not the kind where you sign a contract to work together for a specific period of time, but the kind of post Napster social media sharing that floats beats and mixes out through websites to be downloaded, integrated into live events, shows and clubs, and remixed and re-uploaded for listening and critique. Some websites work on a paid subscription basis, where they keep track of uploads and downloads and DJ's get points for their popularity. In other cases, listeners type comments on the track in real time, noting where the "sick beat" fell.

This collaboration on websites like Newgrounds.com was established in the mid-1990s to create a community that fosters collaboration. Newgrounds section titled "Newgrounds Wiki: About Newgrounds" explains "Feedback is the fuel that keeps us going." When dubstep makers listen to other music they want to contribute. They contribute, collaborate, and receive feedback on their tracks. Listener reviews fuel the circle once again. The ease of sharing and making fuels dubstep. DJ Merc has paid subscriptions to several different sharing sites. He explains:

> There are a couple of DJ websites, one is particular is called yourremix.com, it is a sharing website. I can download as many beats as I want but I have to upload ten original mixes myself. And it was nothing but this DJ website that everyone belonged to and we shared our mixes and downloaded other peoples but they could download from us and at the end of the month they would let you know how many downloads you got, which DJ had the most downloads. You kind of got points for that toward your account. That's one of my favorite websites. I also use a site called DJCity.com. This one is like a blogging website. We still share all different tracks but it is a website where guys will say, O.K., this is making its way down the East Coast. I've had stuff for four-six months before I started to hear it on the radio.

The site yourremix.com has closed down but DJcity is still a viable site to place mixes. DJ Merc likes the abilities to both share his work and incorporate other DJ's mixes. Receiving statistics on his downloads and knowing who was most popular that month was important and it was one of his favorite sites. But what about sharing and incorporating digital music content?

> When I do a DJ gig, I get paid to go out there and do it, and I'm playing other people's tracks with slightly different version of them with a different mix, but I didn't have the rights to that music. So it is kind of hard. It is really hard to say. (DJ Merc)

The creation, the culture and the technology all combine to become complicit in the ethical questions concerning dubstep. The sampling of music a DJ has made is as easy as point and click. Adding beats, or as DJ Merc explains, "a slightly different version with a different mix," allows the ownership water to become muddy. Sharing websites and the feedback and statistical "hit-making" increases the loop of sharing. The repeat ease of uploading and downloading by DJs and fans removes the "original" creator further and further from the mix (literally). At sites like Newgrounds anyone can post, no matter what level of expertise, but the feedback is crucial to improving. Newgrounds has a page on their site called "Collabinator," where a DJ can list his/her talent (an artist, a programmer, a musician, a voice actor, a writer, everyone) who is seeking (the same) as a place to connect in a similar way as the newspaper "wanted ads." Members also list competition information.

Sharing becomes an ethical dilemma that the industry addresses in a variety of different ways. Most sites have copyright clauses and dubstep mixes are often flagged for copyrighted content and blocked from the site. The copyright infringement software for sites like SoundCloud and Youtube have become increasingly sophisticated and the algorithms can "catch" similar beats and times and identify hooks and mixes with copyrighted material in it.

> It is hard to hear anything that is 100 percent original. In all honest, there is an origin in everything. So even if I go out and make a 100 percent original track, it is tough to say that it is mine because I listened to other people, and I put it together to make it my own. So I could be sampling from six different artists, and put it into one, and you could say that it is 100 perecent yours because it was your idea, its your mix, but it was not 100 percent yours because you were clearly inspired by what other people are doing. (DJ Merc)

Sharing also leads to oversharing. Copyright issues have long been a concern in the digital music business.

> That is tricky. Some of the big players might [care]. That is kind of hard to say because it goes back again to the sampling. Whenever I think of copyrights I always think of the Vanilla Ice song that came out, the *Under Pressure* song just made that little sample. But with copyrights it is hard to say. With all the DJ sites out there, it is something that we go into, and we all know, well, if it is an original mix, I know that someone can take it. But that is a risk I take when I log onto that website. But then again it is not really a copyright loss for me because I did not get the rights to the track I am using. So I don't own it.

Most DJ sharing sites have a section that discusses copyright issues, and several note how to get around copyright concerns. A subscription fee is part of most sharing sites, which adds a layer of protection to the idea of membership. SoundCloud's site gives detailed information on the specifics of copyright and allows users to quickly report any copyright infringement. The site provides a checklist and suggests DJs ask themselves questions to specifically orient the site user to copyright concerns. The questions focus on whether or not the DJ composed and wrote the content. Further, the site specifically asks, "Does the track contain the entirety or any part of someone else's song(s). Is it based on someone else's song(s)?" Because, as the site says, "Copyright is complicated." DJcity takes a different route with their site. They explicitly note that the content providers retain the copyright to their content and DJcity is an independent subcontractor to promote the content.

When it comes to dubstep, there is a lack of control for the technology that is used and the way it is used. Sounds that fall within a certain range of sounds are identified as dubstep, even in different cultures. The recognition of the familiar on sites that feature dubstep music multiplies the sounds and the genre and the technology. The exchange is also two-way. The sharing amongst cultures using similar sound technologies both homogenizes and diversifies the sounds and produces splinter genres.

THE MASH-UP

Dubstep has ascended the ranks of popularity and strains of the dubstep nation can be heard wub, wub, wubbing its way through popular culture in music, movies, commercials and the 'net. In the summer of 2012 *Spin* came out with the "Thirty Greatest Dubstep Hits of All Time." Undoubtedly, dubstep has arrived and technology is central to every aspect of the experience. However, the human side of the experience, the talent, is still more important than "the stuff you use." This dubstep investigation has only scratched the surface of understanding the human–technology possibilities of the EDM Makers movement. Technologies like Music Information Retrieval (MIR) and intelligent technologies that use music recommenders; automatic playlist generators and music browsing interfaces are part of the online music store and streaming application experience.

> Learning to hear the unsaid gains entry into this community and history to some degree. The learned is the initiate who has already heard and thus has entered into the community and the history . . . [however] the unsaid can be missing in unlearned listening. (Ihde 2007, 62-63)

The postphenomenological study of digital sound, earbuds, portable sound, and dubstep shows the embeddedness of technological texture. The variations and multistabilities illustrates that sound technologies are artifacts of culture that change the lifeworld in large and small ways. The intent of the study is to recognize that digital media does not lack effect. There is still more to hear. The final group of chapters illustrate that photo manipulation, news and self-tracking textures our world in multiple and varied ways through the postphenomenology framework.[2]

NOTES

1. This case grew from a presentation at Stony Brook University, Manhattan, on 3/21, 2012, and is included in *Technoscience and Postphenomenology: The Manhattan Papers*, eds. Robert P. Crease and Jan Kyrre Berg Olsen Friis, Lanham, MD: Lexington Books, 2015.

2. DJ Merc is the professional name of Tim Mercandetti. He was interviewed for this chapter on 3/20/12. Portions of his transcribed interview are shared here and he signed a permission for use in this book.

Chapter Nine

Case: The Photo Manipulation Aesthetic

Stop for a moment and take a look around you. Nearly every image you see today—in ads, on billboards, in magazines, on websites and in newspapers—was touched in some way by photo manipulation tools. Its influence is so great that one program has even earned a place in the vernacular: The verb "to photoshop" has become shorthand for the act of altering digital images (Pfilffer, 2010). "For humans, there can be no god perspective, only variations upon embodied perspectives" (Ihde 2002, 70). Digital media, as content extends and morphs through technological instruments, become artifacts in their own right. This chapter 9 case shifts from aural digital media to the visual example of photo manipulation software to continue to illustrate the human-technology connection.

Photo manipulation has occurred since photography was invented in the mid 1800s. But early photo manipulation, as creative and cunningly "invisible" as it might have been, is no match for today's digital image editing tools. A look at photo manipulation widens the digital field to contemplate the aesthetic of the touch-up and beyond. Trick photography has been around for a long time. Dino A. Brugioni's 1999 work *Photo Fakery: A History of Deception and Manipulation* comprehensively describes the history of photo manipulation. One of the most popular and current image manipulation processes is called "photoshopping." The name of the trend, and a viable career skill in many industries, comes from the software most popularly associated with it, Adobe Photoshop™. Digital manipulation software is close to twenty-five years old, and other kinds of photo manipulation software programs and applications have been cropping up ever since. In a techno-utopian way, the tool users make the body image "perfect." With today's technology the "user" can be a thirteen-year-old girl touching up blemishes on photos before she uploads them onto a social media site, or a graphic designer preparing a photo spread for a million dollar magazine (Harrison and Hefner, 2014; Adatto, 2008).

The newly constructed image comes with a constructed perception situated squarely between the actual body and the digitized or imaged body. This case also studies photo manipulation through the postphenomenological lens. A look at the trajectory of the variations of the manipulated image says much about a socio-cultural shift that takes on a different name for "real," in the pursuit of perfection, and an understanding about what it means to be human.

EXPLORING THE SOFTWARE

In the fashion industry, the process of image editing can involve but is not limited to manipulation that cinches a model's waist, adds a gap between the thighs, changes hairstyles, reconstructs facial features, erases wrinkles, enhances clothing drapes, adds muscles, shortens skirts, enhances muscles, extends legs and changes skin color. The photo manipulation process is also called photo retouching, airbrushing, editing, photoshopping and, on the extreme end, "photo-cleansing." The ability to alter has become an increasingly popular topic in mainstream media, news, college campuses and academic study. The politics of digital photo manipulation are far reaching, and these altered images are having an effect on culture (Yamamiya et al, 2005; Hawkings et al, 2004; Agliata et al, 2004; Holmstrom, 2004; Magee, 2012). Photo manipulation produces variations of photo images of women and children in fashion magazines, catalogs and social networking sites. Adobe Photoshop software costs less than $200.00. Photoshop also now has competition with open source programs like GIMP (GNU Image Manipulation Program) and Adobe's web application, Fireworks that also edit images through a variety of additional features optimized for niche markets. Tablets and phones can also access less expensive apps that do similar touch-up work like filtering and editing.

The process of photo manipulation codetermines the constitution of subjectivity and alters the reality of the lifeworld in subtle and not-so-subtle ways at a time when doctors around the world are publishing studies that confirm the negative effects from the hyper-sexualization of young girls, and economists note the rise in spending for wrinkle cream and cosmetic surgery. Studies show that close to 100 percent of all published images in the mainstream media are photoshopped, and the majority of social media photos have used image editing. Free shareware can also be used to check to see if the pixels in a photo have been altered. Digital forensics experts say that photo manipulation can be detected through checking the file and metadata, looking at shadows and reflections, and noticing even microscopic traces of tool marks from the software. The altered image of the body "mediates" the connection between the viewers and the lifeworld. The subjectivity and objectivity of the technologically mediated intentionality shifts. The postphenomenological framework probes for multiple variations that constitute the multistability of photo-manipulated images.

I have been challenged to think about what it means to go "to the cloud" (storage) to switch the heads a family member to create a more aesthetic image to send to loved ones. I began to wonder, what is the interest in perfecting the image? And how does an aesthetic, as a set of principles that guide the appreciation of beauty and of a set of standards for photography and portraiture, change through the use of technology designed to alter. What might it mean for a six-year-old to know that her dad changed her head in the family photo, or for an adolescent to see her mother touching up a photo before it goes onto social media? What kinds of things do photo-shopped images bring forward about the human body as an object to be tweaked and fundamentally altered for the gaze? A quick Internet search can produce many articles and examples of photo manipulated images and memes. In 2009, a photo manipulated image of singer Kelly Clarkson appeared on the front page of *Self* magazine. Clarkson's body had clearly and obviously been altered, and fans noticed and cared. The Internet lit up with blogger comments, which gave *Self* editor Lucy Danzinger an opportunity to discuss photo manipulation and the philosophy behind it in the fashion magazine world.

> Did we alter her appearance? Only to make her look her *personal best*.... But in the sense that Kelly is the picture of confidence, and she truly is, then I think this photo is the truest we have ever put out there on the newsstand. (Redefining Beauty)

So what can be said for the personal best image—personal best or impersonal fake? Some photo-editing professionals note that as soon as a photographer alters the light situation, the image has been manipulated. What might be the difference between physical manipulation through perfectly applied make up, draped clothing, coifed hair, expertly set lighting, camera filtering and lens manipulation by an expert photographer and manipulating an image post photo shoot? In a 2009 *New York Times* article, Randy Cohen explored the ethics of photo manipulation, which is gaining steam in all industries. He wrote that professionals now have standards and codes about what can and can't, should and shouldn't be manipulated, touched or otherwise altered, but decisions are company and country specific. There seem to be no far reaching or overarching consensus on the amount and kinds of photo manipulation a company can do, from a legal standpoint.

Some professional media organizations, like the National Photographers Association, have a "Code of Ethics Statement of Principle" that explains "Editing should maintain the integrity of the photographic images' content and context. Do not manipulate images or add or alter sound in any way that can mislead viewers or misrepresent subjects." In June 2011 the American Medical Association (AMA) adopted a new policy that states, "The appearance of advertisements with extremely altered models can create unrealistic expectations of appropriate body image. In one image, a model's waist was slimmed so severely, her head appeared to be wider than her waist," said AMA Board member Dr. McAneny. "We must stop exposing impressionable children and teenagers to advertisements portraying models with body types only attainable with the help of photo editing software" (American Medical Association, 2011). Britain, France and Israel have considered legislation to create guidelines for the amount of digital photo manipulation that can be done on an advertisement, along with a disclosure that states the image has been altered. In 2011, the National Advertising Division of the Council of Better Business Bureaus in the U.S. ruled an ad for a mascara-featuring singer Taylor Swift was misleading and the manufacturer permanently discontinued the advertisement. Worldwide, exaggerated and overly airbrushed images have been banned on the grounds of being misleading (Neff, 2011).

The conversation sweeps deep and wide, from celebrities and models to politics and sport. Celebrities often do not comment on airbrushing because they need the media's approval and do not want to make waves about photo manipulation. On the other hand, some famous individuals tweet their own "selfies" after they personally touched up the image. Britain banned L'Oreal hair color ads because they were airbrushed and gave the notion of unrealistic expectations about the products, thereby constituting misleading the public. A 2009 *New York Times* article noted a comment from a Paris-based photographer, "I have never yet seen, and you probably never will see, a fashion or beauty picture that hasn't been retouched (Pfanner, 2009)." On their website, the Association of Magazine Media notes that 99.9 percent of fashion magazine photos have been altered in some way, from blemish retouching to body alteration, to skin tone lightening or darkening.

Postphenomenology's key concept of multistability can be used to further understand the phenomenon of photo manipulation. As noted by Ihde and Verbeek, technologies have a certain directionality, a trajectory that shapes the way they are used. Digital-alteration software has tools to brush, sharpen, crop, touch up and erase a digital image. The tool palette has buttons like spot healing, a healing brush or patch tool to quickly remove imperfections, and "clone stamping" allows the user to repeat the same element over and over. The toolbar elements create options that explore a certain directionality. An "essentials" panel provides colors, swatches and styles through layers, channels, paths and panels for fixing and masking an image. Adobe provides a variety of online tutorials for touching up skin, face, eye, nose, hair, teeth, body and hands to make a photo flawless and perfect. The software mediates the connection between the user and the digital image. The software, by virtue of how it is designed, labeled, marketed and taught, plays a role in the context for its use. Photo manipulation fixes things. Fixing suggests brokenness. The "fix" can span from subtle to extreme, but the software is not a neutral intermediary "between humans and world, but mediators: they actively mediate the relation" (Verbeek 2005, 114).

VARIATIONS ON A THEME

One way to study the variations of a photo-manipulated image is to explore a series of variations. This series starts within an untouched photo (figure 9.1) of a girl in her neighborhood. She is naturally attractive and standing outside, looking at the camera.

Figure 9.1.   Unmanipulated image

Variation #1: The Touch Up

Figure 9.2 is a picture of the same girl, but some components of the digital image have been altered. Her teeth, hair color, complexion and eye color have been "enhanced." Variation 1 shows someone who looks naturally at-

Figure 9.2.   Variation #1: The Touch Up

tractive, and if you saw this photo without the original you might not know that there was an alteration at all. In many cases, this is the level of photo manipulation that occurs for most "touch ups" on social media postings. Also, within the context of news or fashion magazine images, a celebrity's photo will be altered in a photo spread or story. Slight photo manipulation might be used to enhance the celebrity's features. The image is slightly "touched up." Blemishes, dark shadows, wrinkles, and odd clothing folds are removed. Eyes might be slightly altered to darken the color and teeth might be brightened. The touch ups may or may not be noticed, but the image presents a more perfected and stylized look. The viewer is focused on the looks, talents or celebrity activity illustrated in the photo layout. In general, the images altered to the level of the first variation are done to enhance the beauty that is there and to remove noticeable blemishes, clothing wrinkles and awkward lines. In many cases, the image is seen as normal for that type of magazine. The perspective or embodiment is similar, a human body, but altered based on clothing, position, action and context. The same, but perfected through photo manipulation. The variation presents a slightly augmented reality.

Variation #2: Transhuman Form

A transhuman form resembles a human form in most ways. The Swedish philosopher Nick Bostrom in his article "A History of Transhumanist Thought," explains this re-assembling as the possibility and desirability of fundamentally transforming the "human condition by developing and making widely available technologies to eliminate aging and to greatly enhance human intellectual, physical, and psychological capacities" (2005). Transhumanist thinkers study the utopian and dystopian sides of emerging technologies that could overcome fundamental human limitations moving toward another variation. Certainly one's feelings of embodiment—implied perceptual bodily position and satisfaction with ones own body—feeds into the variation. This photo manipulation goes one step further by refashioning the image into perfect ideals. The shift in thinking goes from perfecting the current features and removing slight blemishes, to redefining shapes and features toward a "perfect" prototype. Figure 9.3 shows the same girl, but her hair has changed color, her face is noticeably thinner, her lips are fuller and more deeply colored red and her eyes are a brown color. Her hair is also now a brighter red. Anyone who knows her would notice that the image is more

Figure 9.3. Variation #2: Transhuman Form

than a "touch-up" if they saw it. In many cases, transhuman figures are bodies that have been enhanced to break away from one's current body limitations. In the world of photo manipulation, this means enhancing and lengthening current features. The humanity in the variation is transformed beyond normal boundaries toward ultimate perfection in all categories in the image. Eye and hair color can change, cheek bones and nose can be altered, eye distance may be diminished or widened, hair might be changed, neck might be lengthened, waist tightened, legs lengthened and the overall physiology of the person in the image might be augmented in ways that move from a "touch up" to an altered being. The person reflected in the image is no longer someone real. The viewer suspects the image is different or altered, but buys into the idea that people who look this way do exist somewhere, but the viewer is not one of them. The image mirrors a cultural idea of bodily perfection, without moving into the inhuman form.

Variation #3: Inhuman Form

Figure 9.4 explores an image that is altered to the point of not looking like the original human it was modeled after. It is aesthetically outside the boun-

Figure 9.4.   Variation #3: Inhuman Form

daries of what we expect when we might look back at the original model. The variation does not harken back to the human form it was originally made from. These changes can be both subtle and overt. We see the point of view and it is radically altered. We see the other in the image but the qualities are quite different. The eyes do not look quite human, and the spark has left the eyes, the skin looks so flawless and shiny it seems plastic, and the body somehow, for some specific reason or from poor photo manipulation skills, does not look human. It looks visibly altered. The inhuman form does not look like a cyborg or a transhuman, but inhuman in scale, with a glassy-eyed gaze, or altered or decomposed features. Jacque Lynn Foltyn, in her chapter, "Corpse Chic: Dead Models and Living Corpses in Fashion Photography," explains:

> Modeling "death" to sell fashion is an increasingly common motif. In the editorial and advertising pages of *Vogue*, *W*, and even *The New Yorker*, supermodels, actors, and those famous for being famous are styled and photographed to look freshly or long dead as they pose for pantomimes of literature (*Romeo and Juliet)*; music (*Death and the Maiden*); and film, TV, and popular fiction genres (*film noir*, vampire, CSI). (2011, 380)

Sometimes an advertising agency will manipulate a photo until the image takes on decidedly animal forms of a cat or a wolf. Other times, the image can look soulless or illustrate a loss of connectedness in the eyes. Adds Foltyn, "The current fetishism of death as beauty that is apparent in fashion magazines has a ghoulish logic. Fashion models are sometimes referred to as paper dolls; a paper doll is static but can be posed, just as a dead body is static but can be posed" (386). And most, but not all viewers intellectually know that manipulation has occurred. The inhuman photo is too "othering" to be self—a distinct alteration. Not human but close to it, with other forms juxtaposed onto it. In most cases, the inhuman form looks dead, with vacant eyes and a dislocated form with a body and facial features that are not quite proportionate.

Variation #4: Posthuman Form

Figure 9.5 is an illustration of a speculative being that represents or seeks to enact a re-writing of what is generally thought of as human and goes far beyond trans-human and inhuman, as in the senescent and extremely intelligent beings of science fiction work. The form is beyond human to the point that the viewer might think of a robot at first glance, and then move toward a consideration that the form may be human. In general, this variation, when seen in fashion magazines, takes on a decidedly robotic quality. The gaze seems to be more than just vacant or "dead," but unconnecting in an otherworldly way. The stance and expression are often stiff and there is a com-

Figure 9.5. Variation #4: Posthuman Form

plete lack of emotion in the facial features. The viewer easily knows the distinction as a non-human form with human features. The alteration has moved past human to the point where it does not seem like the image started with a human image at all. In many cases these images are not overtly robotic, like transhuman forms that illustrates body parts made of computers or prosthetics, or skin made of metal. The posthuman form, illustrated through images, is becoming other. The trend is currently seen in super heroes and video game figures, who combine the human form and anime to move past human movement—but not toward looking cyborgian or robotic, but embodying artificial intelligent aesthetics, super-human bodily processes, or alien/zombie embodiment (Crease, 2006).

PARSING THE PARTS

Active perceptual engagement, the ability to situate oneself within the image, is an important part of a viewer's image analysis. Is the viewer engaged or a spectator? In a certain environment, like a fashion magazine, the experiences with these images, compounded by cultural norms, the text that attends or does not attend to the image, and the material world around the person engaged with the image, specifically situate the viewer in the lifeworld. Technologies bring new possibilities that change the context of the role of an image.

But really, as I was told one time, they are just images. People have been touching up senior pictures and school portraits for years. Isn't it the same thing? In the same way that radiologists need to upgrade their "viewing" skills because new techniques are always being developed, so too must viewers of fashion magazines and advertisements upgrade viewing skills (Prasad 2005). During MRI radiological analysis, images of only a small part of the body are produced. The physician decides which part of the body has to be imaged on the basis of her or his initial diagnosis of the patient's ailments. So, too, in fashion magazines, do legs, lips, hands, faces, arms, torsos and eyes become dismembered and objectified, produced images for analysis. As a comparison, the work of the MRI technician's reading of an image is not unlike the work of a consumer reading an image.

For the MRI technologist, the process is not automation. The viewing is a kind of seeking for information. So too is the gaze of the young, middle-aged or older person, who seeks information or just happen to want to read a magazine, but finds that the gaze upon the photoshopped body is everywhere—human, transhuman, inhuman, posthuman. Flawless, wrinkleless, blemish-free, toned and shiny, tanned and tweaked. MRI images produce different reconfigurations of the body, each of which provides a partial perspective and together they constitute the MRI radiological gaze. Manipulated photos produce different reconfigurations of the body, each of which provides a partial perspective toward a particular mediated gaze. As Sociologist Judy Wajcman reminds us: "Technology is more than a set of physical objects or artifacts. It also fundamentally embodies a culture or set of social relations made up of certain sorts of knowledge, beliefs, desires, and practices" (1991, 149).

Anne Balsamo, in her article *On the Cutting Edge* notes, "Carole Spitzacks suggests that cosmetic surgery actually deploys three overlapping mechanisms of cultural control: inscription, surveillance, and confession" (2002, 686). According to Spitzack, the physician's clinical eye functions like Foucault's medical gaze; it is a disciplinary gaze, situated within apparatuses of power and knowledge, that constructs the female figure as pathological, excessive, unruly and potentially threatening. The gaze disciplines the unruly female body by first fragmenting it into isolated parts—face, hair, legs, breasts—and then redefining those parts as inherently "flawed" and pathological. When women internalize a fragmented body image and accept its "flawed" identity, each part of the body then becomes a site for the "fixing" of her physical abnormality. Spitzack characterizes the acceptance as a form of confession. "In that same way that cosmetic surgery is not then simply a discursive site for the 'construction of images of women,' but in actuality, a material site at which the physical female body is surgically dissected, stretched, carved, and reconstructed according to cultural and eminently ideological standards of physical appearance" (2002, 687). The photo manipulated image is a discursive site for the construction of images of women—and men—digitally dissected, shrunk, cut, tweaked, smoothed, reconstructed and otherwise tooled according to cultural and ideological standards of physical appearance. The radio journalist Esther Honig embarked on an interesting project of discovery she called "Before and After." She took a self-portrait and then sent it around the world to be manipulated by graphic designers. Honig explains:

> In the U.S., Photoshop has become a symbol of our society's unobtainable standards for beauty. My project, "Before and After," examines how these standards vary across cultures on a global level. Freelancing platforms, like Fiverr, have allowed me to contract nearly forty individuals, from more than twenty-five countries such as Sri Lanka, Ukraine, the Philippines, and Kenya. Some are experts in their field, others are purely amateur. With a cost ranging from five to thirty dollars, and the hope that each designer will pull from their personal and cultural constructs of beauty to enhance my unaltered image, all I request is that they 'make me beautiful' (Beauty Standards around the World, 2014).

Each of the photos that she shares on her website, reflects both the cultural understanding of beauty and the skills of each of the photo manipulators.

Technological reconstruction, through a software program and a computer, digitizes an image and refashions a more "perfect" one for the gaze. But who defines perfect? The case of photo manipulation and its variations, which by no means have been exhausted here, results in changed images in social, cultural and personal life. Multistability and variation in postphenomenology have highlighted a way of visualizing the potentially invisible, and photos are not the only kinds of illustrations altered by software programs and digital manipulation. The last four cases continue to illustrate the ways digital media has technologically textured the lifeworld. Chapter 10 studies data mining technologies. Chapter 11 explores aggregate news, and chapter 12 studies self-tracking experiences that connect humans with technologies in new and interesting trajectories.[1]

NOTE

1. An earlier exploration of this topic occurred in a presentation shared at the Society of Phenomenology and the Human Sciences (SPHS) Conference in October 2011.

Chapter Ten

Case: Data Mining

As the cases have illustrated thus far, technology is not neutral, and digital media technologies are engaging and transparent in their use. But there are interesting differences that open up each technological artifact in various ways to increase understanding of our human–technology connection. Peter-Paul Verbeek's essay *Artifacts and Attachment: A Post-Script Philosophy of Mediation* (2005b) explains "Technology should be analyzed not only in terms of social processes in which it is constructed, but also in terms of the role it plays in social processes itself," (125). This statement can be no truer than in the examination of processes and technologies digital makers use to craft journalistic news stories. Chapter 10's case focuses on variations and multistabilities of mining data for media consumption[1] and chapter 11's case delves into the multistabilities of news content.

ARTIFACTUAL

It is important to first think about the idea of data-mining technology as an artifact. Cathrine Hasse, in her article *Postphenomenology: Learning Cultural Perception in Science,* pursues an artifact in a way that can be helpful in interpreting some of the variations and multistabilities of data-mining technologies.

> Artifacts are not just material entities, they are also signs—and it is as signs (or pictures or pictograms) that they can be psychologically internalized. It follows that artifacts from the very outset are imbued with cultural meanings, which are learned and internalized not just in use, but also in social use. (2008, 48)

The cultural psychological perspective she uses pushes the idea that artifacts are created by humans—software designers, engineers, computer programmers—with purposes. Artifacts can be assumed and adopted into culture and society in such a way as to be taken for granted, or accepted without scrutiny. She adds, "They are tools meant to fulfill a desire or requirement developed in a cultural group of people, who pass not only material artifacts but also their cultural meaning on to the next generations through a process of learning" (48). So what can be learned by analyzing the use of data mining as a tool in journalism? What might be internalized or taken for granted by journalistic artifacts in the form of digital media and the associated technological devices? And what kinds of co-shaping or co-constitution occur in this human–technology connection? Hasse's assessment is that, "[i]n a postphenomenological perspective objects become multistable and new questions can be asked in line with the problematization of representationalism" (2008, 57). This is a reminder that the "artifact" consumed by the user is artificial in nature and represents the content in a specific way, a representational way. How is news and information represented, and—re-presented?

> Artifacts mediate ways of existence (subjectivities) and experienced realities (objectivities) not because people told them to do so, but because of the relation between humans and the world that comes about through them. This artificial representation, while being the best content that we might get on a subject in a quick and interesting way, is still a representation. (Verbeek 2005, 140)

One of the interesting things about digital media is the way the industry fosters and multiplies itself. Television talk shows create content about newly released films. Film producers are interviewed on TV shows. Radio shows discuss the top picks for mobile phones for the year, and they all review the top music, the top games and the best apps to buy. Media proliferates and re-presents the same parts of its content over and over. They broadcast or 'net cast award shows about the content they create. And they all compile statistical data about each other, which they analyze on an almost-daily basis. The contemporary media environment allows everyone to be an aggregator and curator of content. Have hyperlink will travel. The data-heavy analytics occur for both the programming and the underlying business model. Social media puts the content and the analytics about the content into everyone's hands to follow and share. Providing this information fosters increased use. And use creates data. And data can be reported as the media continues to create content by and for *itself*, in a simulacratic way. Digital media users often see carefully embedded numbers as facts (infographics) and everyone loves cold hard facts in a world that seems increasingly complicated and difficult to figure out. Users can check out their personal analytics, be they large or small, on almost every digital media platform available. Organizations and individuals scrutinize likes, views and hits along with the geographic location and duration of the data traffic. Digital media provides feedback much more quickly than the tools used in earlier days.

In *Big Data Now: Current Perspectives from O'Reilly Media* (2012), *data* is explained as having "valuable patterns and information, previously hidden because of the amount of work required to extract them" (ch. 2). So what kinds of data are extracted and collected?

> Input data to big data systems could be chatter from social networks, web server logs, traffic flow sensors, satellite imagery, broadcast audio streams, bank transactions, mp3s of rock music, the content of webpages, scans of government documents, GPS trails, telemetry from automobiles, financial market data, the list goes on. (ch. 2)

One illustration might be triangulating data from social media, geographical information and shopping receipts to inform ideas about peer influences in shopping decisions. Statistical analysis has always been part of the media world. Advertisers have always wanted to know who was reading, or listening or watching so their advertising dollars and content were available to the largest media population they can afford to share it with. Media personalities have always wanted to keep their profiles high to garner the largest market share of popularity.

At my first radio station job, everyone knew what day the Arbitron rating "book" came out. Even when media was primarily analog, the stakeholders lived by viewing statistics. The two most widely known ratings companies, Arbitron and Neilson, capture a representative sample of how many people listen or watch media in their homes, times they watch, their geographic region, viewer demographics, and a variety of other data depending on the kind of service a media organization is using. The "capture" occurred through written diaries and then later on, electronic meters. Measuring log-ons, downloads and activities have made statistical analysis much easier because information occurs instantly. Media analysts can watch program statistics in real time in testing facilities, and focus groups help owners and financiers gauge popularity for advertising revenues. Many media organizations link promos with websites to track interested listeners, create contests to track interest, open discussion boards for feedback, and have connected online radio stations with terrestrial radio stations to coordinate efforts for measurement. Suffice to say, statistics make the media world go 'round. In the world of journalism, the statistics are called data.

Chapter 10

DATA-TUDE

The etymology of *data* suggests the word stems from the Latin word *datum*, or thing, given or "to give." More current meanings of computer information and data processing are used for the first time in the 1940s and 1950s. So data is a thing, given. Data-driven or data-centric journalism took hold when reporters and producers started to use computers to assist with story analysis in the later quarter of the twentieth century. Recent trends for data collection include quantified self-tracking devices, data visualization mapping, data libraries, data coding, and sensor data from electronics. When the international online organization Wikileaks went mainstream in the summer and fall of 2010, to publicized a compilation of close to 500,000 documents related to gun camera footage from a 2012 Baghdad airstrike, the idea of data mining, sometimes called knowledge discovery, came to the forefront as a journalistic technique. Part of the journalist's job has always been to "dig up" the story, so hardcore investigation has always been part of some journalist's job descriptions.

MINING IT

There are five basic definitions of the word *mine*, including the Old English pronoun mine, from *my* and several different noun roots. A mine can be a pit or tunnel for obtaining metals, first seen in the 1300s, and the explosive variety—to dig under foundations to undermine in the later 1400s. The notion of digging, mining and extermination were used from the mid-fourteenth century, as to dig in the earth for treasure. Roots of the etymology of data mining suggest giving something that is dug from a pit or tunnel from underground, sometimes to find explosive or undermining information. In data mining, the information is the commodity.

Information has long been the focus of media types around the world, but in some ways, data mining is different. Some journalists are specifically called data journalists. In recent years the International Data Journalism award has been created to award computer-assisted reporting for a job well done. Awards aim to set high standards and best practices for data journalism. Past awards have been given for data-driven investigations, interactive data visualizations/info graphics and data-driven mobile and web application services.

The metaphor of traditional mining helps to better understand the data-mining process. Both mining and data mining start with prospecting. A mining prospector spends a lot of money and time looking for a good mine, hoping to find something. He or she purchases the equipment and heads off in a particular direction, hoping to keep the location a secret so others do not find out and head to the same place. A miner's tools included picks, shovels, hand drills and blasting powder. Some mining towns have big machinery at work drilling for ore, while others used hand tools specifically focused in on small valuable veins of precious metals. The deeper miners dig, the more dangerous the work becomes. Mining a database for information in stored scanner data in a large database takes on similar techniques to mining a mountain for a vein of treasured ore. Sifting through data and probing for linkages in the vast dearth of digital dirt is tedious and often painstakingly mundane. But data-mining technology has become one of the newest forms of journalistic newsgathering. This kind of discovery can supplement current journalistic investigations or find data for new stories to report in a wide array of new and traditional media. And every now and then you strike gold.

Data tools like data mining, data manipulation and data visualization, are all part of the storytelling process. Mobile sensors, social media, video surveillance, video rendering, smart grids, geophysical exploration, medical imaging, and gene sequencing all provide data creation. Digital tools loaded with analysis software big and small dig deep into the analytics. Much of the work is in text analysis that comes from unstructured information, like emails and call center transcripts, or semi-structured sources, like server logs or social network API reference documents. Parsing, searching, retrieval and text mining can discover meaningful insights, and additional clustering and classification strengthens the analysis. Journalists who work with big data, along with data scientists and data engineers, progress through six phases of data analytics lifecycle to discover where data might be located, prep the data, plan and build a model to discover the relationship of variables, communicate information and act based on the knowledge gained.

Computer analytics can predict hot spots of data use. Journalists pay to learn this information to keep on top of hot stories. Sometimes the break comes through stream monitoring—what's trending on Twitter, for instance. In some ways the news of today still carries that historical banner of getting the story from people who don't want to give it. The tools are different, and some reporters use large computers while others use small devices, but journalists who used to have to get to the inner circle to get invited to the right press conference now use the newest technology to dig for the story in a different way.

Developments in digital media and mobile technology continue have a dramatic impact on journalism, in the way consumers get their news and in the way journalists report it. As mobile devices promise "TV anywhere, anytime," consumers make use of greater connectivity across social networks. The pressure is on professional journalists to respond and act in similar ways, using similar technologies. Citizen journalists have gained traction for breaking stories because they are out and about the lifeworld, devices ready. The professionally paid journalist now has to try that much harder to "get the scoop." Additionally, citizen journalists are receiving journalist protections for the work they do on their digital-media device (Gant, 2011). Court cases involving the right to privacy, the right to reveal sources and the idea that a digital device is private property, are gaining worldwide traction. The idea that the police can search a professional journalist or citizen journalist's cell phone, which might reveal sources and additional information, has led to debates about Fourth Amendment rights in the United States and legal debates in other countries. The use of data and high-end software programs to crunch data and pull together resources is one of the ways that established media organizations and their journalists can remain relevant and competitive while reporting the best story.

Data mining has become a technologically mediated place in the co-construction of our objective and subjective world. Data-mining techniques use software to create decision trees, graphical inferences, visualizing clusters, nearest neighbor historical classifications, artificial neural networks, genetic algorithms and rule induction. The techniques are part journalistic newsgathering and part computer assisted reporting. In many cases, data-mining work occurs within a team of journalists, designers and programmers. A postphenomenology study of data mining reveals technological mediation, co-constructive technological intentionality, and multistability, to reveal variations to the artifact called data.

TECHNOLOGICAL MEDIATION

Data mining mediates the data in the world. The journalist or researcher, the web, the data, the computer itself and the sophisticated software programs co-construct the material product of data mining, the final output. In many cases this comes in the form of an image.

If seeing is believing, then seeing statistics in a readable form like a bar graph or infographic is doubly believable—especially after the data has been cleansed of any extra fragments or data parcels that distract the specific intent of the story. At its core, data mining is a technology that is very similar to medical image technologies. In a pragmatic way, data mining is designed to portray the interior of data that is distributed throughout the world and housed in data warehouses and clouds. In the same way that medical imaging makes it possible to perceive the inconceivable, data mining uses pattern recognition and gestalt theory to created graphic visualizations. The plot lines are data available in the world, instead of within our bodies. What is revealed and concealed in the development, in the information itself, and in the reading and cleansing of the image? In general, journalists have not had access to this kind of data before, because it has been too vast. Data mining allows for an image to appear. The newly imaged data is presented and makes information and ideas newly apparent. Ihde notes, "the medium (the data representation), contains in its realism/irrealism presentation an effect in which 'image technologies' mediated and unmediated ordinary experience against becoming dialectical," (1993, 49). We cannot forget the data-mining process and our part in it. The process of how our news arrives to our digital-media device should not be taken for granted. This is a non-neutral impact involving compound seeing. The infographic magnifies potentially mundane statistics and facts, processes them, and makes them contextually meaningful for media consumption.

Data mining co-shapes the way media consumers view data. The data artifact re-presents the statistics and their cultural meaning. And for many, reading media content or data, scraped, cleaned and visualized in a tangible way, makes one feel more informed, media literate and more educated. The technological bundle—journalist + device + web + interface+ data—directs the human–technology experience. This trajectory began way back in the history of media. But the "tools of the trade" are changing. The discipline of journalistic research is changing. The technology package or bundle—mediates the data and makes a new way of seeing information possible.

We, as users who share our life information, are participants in creating an environment that fosters data mining. We share our political views on *Buzzfeed* polls, we answer quizzes about the kind of shoes we buy and troll the web for coupon codes. This provides the data that is used in data mining. Journalists use high-powered technology—software and digital storage—to access and correlate the data to find trends, which often result in data visualizations for media. Web beacons, cookies, scripts, pixel tags and other tracking technologies are used on almost every site from *Huffington Post* to *Thought Catalog*. Data mining mediates the relation between media consumers and the informational world available on the web. As Verbeek nicely puts it, "When technologies mediate the intentional relation between humans and world, this always means, from a phenomenological perspective, that they codetermine how subjectivity and objectivity are constituted" (2010, 116). This co-shaping artifact, called the data-mined image, can influence humans and the world, but they can only be understood in relation to the human use of them. The data-mining artifact has no relation to the person who does not see the information. There needs to be a perception of co-shaping. This leads us to multistability. The technological process of data mining, as well as the resulting data or image, is both deconstructive and structural, alongside, as Ihde said "our multiple, refracted and compound perspectives" (1993, 87). This is why postphenomenology is a fruitful way to explain data mining. The connection is context dependent. Focusing on trajectories, variations and multistabilities creates opportunities to think differently about data mining.

TRAJECTORIES OF USE

Data mining, as a technological process, changes the information we have access to and the way we receive that information. It alters our human–technology connection. Ihde explains: "The human–technology interaction is one that allows for different trajectories of use, for different possibilities, those that are clearly *non-neutral* but also short of anything like a determinism. And the change in technologies produces changes in what and how ideas are communicated . . ." (2009, 75). Data mining, as a journalistic tool and for larger technologically mediation aims, has different trajectories of use for different possibilities. And the use of digital media content and devices to consume journalistic reports created through data mining, create non-neutral content and instrumentally mediated experiences for users. Once the data is published, pushed (or leaked), users access it through a variety of locations on their device or other technology. For instance, data mined information is released by paid journalists, citizen journalists, public relations specialists, social media users, government officials, law enforcement, and just about anyone who has access to data and a way to crunch it, spin it, and report it. The information is seen on aggregate news sites, social media, and websites of all kinds. The information is accessed through technology big and small, on monitors, billboards, computers, digital display boards and digital media devices.

MULTISTABILITY

What are the essential structures of data mining? What multistabilities can be discovered? At first, mining underground, in tunnels or pits, for things, seems like a labored, dirty and manual process, a far cry from the ways journalists might research a story. What are the parallels between mining and data mining? What are the differences? Upon surface consideration, it may seem that the processes are very different, but you can exchange a few key words and see more similarities than differences. In Don Ihde's 2009 text, *Postphenomenology and Technoscience*, he analyzes archery. He takes us through the bodily engagement from the English longbow, to the Mongolian horse bow, and on to the Chinese artillery bow. The socio-cultural aspects of his multistable analysis showed me the cultural context of the practice of extraction, and the increased training or practice we have in giving, using, receiving and creating useful data. In fact, statistically, the programmers know that we'd rather store the information than retype it so we download those benign little data bits with the cute name—cookies, so it is easier to feed the data mine. The environmental model has not changed. What might be the technology—the picks, shovels and hand drills of the journalist—

digging for the story? What tunnels do they pursue? Various data retrieval and submission tools may be found on the internet. Sometimes these databases are called data warehouses. They can be library portals or city government statistics. The point is to extract hidden but predictive information from large databases. Not so different from mining ore, silver or coal.

When every digital media click produces data, the content piles up. Structured data is the easiest to mine, but tools created for unstructured data are increasing in availability and decreasing in price.

> In fact, one need make little effort to hide one's self on the Web, as anonymity lays claim to the subject. Everyday browsing, reading, playing, shopping, and other activities are logged and accumulated in a data store. Social networks and other hardware/software systems record (and sometimes broadcast) the whereabouts of users, proposing that the availability of the data will further social interaction. In aggregate, these data constitute a new currency, as though wisdom or at least profitable knowledge will accrue to those who leverage them properly. Each participant contributes value simply by going about her ordinary activities. (Evens, 2012)

Data collection has been part of the modern world for many years. File cabinets, closets, boxes and attic storage stacks were typed into computers, which in turn multiplied hard drives and file data. Digital storage became an issue, and large storage mainframes were invented and more frequently used. The most recent shift has been to cloud storage, where the data is shifted to an off-site storage facility owned by a third-party company that "manages" the storage abilities and the multiple data servers. A user uploads, or "backs up," information to a server on demand or during a specifically scheduled time each day, each week or when needed. Some of this data is described as "big data" because the group is too large or too complex to easily break down into usable information. Data can optimize or verify information, identify key information, isolate documentation and predict opportunity and scenarios. One of the trends in data mining is to find ways to use this data, often collected for many years, for investigative journalism to aid in storytelling.

Whether it is census material or votes or who purchased the most products, people like to keep statistics, at least to know what was popular, to purchase more and to make more money. The Excel spreadsheet became the point of entry for data mining, in many cases. The more people transferred to Excel, the more the files could be merged, purchased and become the root of relational databases using a structured query language for acquisition. The more information that is digitized into those relational databases, the more data is available for the warehouses. The more information becomes digital, the easier data mining becomes. Websites like ancestry.com have provided a variety of valuable historical documentation about someone's ancestry as historical data is digitized and compiled for the future. Data warehouses hold content predictors of past information and can produce future trends through the use of advanced algorithms, multiprocessor computers and massive linked databases. Text Term Searching in databases that have been created by governments and private enterprise has also become an interesting and reliable way to gain information for news stories. All statistical materials, from government documentation to social media information, can be text term searched for a price. Data miners can get information about who calls whom, how long they are on the phone, and whether a line is used for fax as well as voice; uncover hidden links between illnesses and known drug treatments; and spot trends that help pinpoint which drugs are the most effective for what types of patients, or what service customers would pay for versus those they felt they should get for free.

One of the ongoing issues of data mining involves information privacy. Companies have a variety of privacy policies that remind users how information is used. However, in many cases, use, in general, generates data that is saved. Any purchase made using a digital-media device is saved somewhere. Privacy policies ensure that personal information is not saved, but user trends are often compiled and when ownership of media companies change, the information travels to the new owner, too. Companies track and compile data on purchases, travel, medical appointments and procedures and employment applications. School records, warranty records and employment records are compiled and cross-referenced, depending on who owns what data. As the data grows and the software programs become more accessible, the ability to connect the dots increases. Data miners have cross-referenced data to make changes in public policy, track financial records and spending, and predict travel trends. Data forensics is one of the up and coming careers in the data-mining field. Adding GPS locations to data allows journalists to track the movement of crime rings and fraudulent activities. Knowing how to cull data for information has become a priority for stakeholders on both sides of law enforcement, and knowing how to access data has become an important new job skill with sweeping impact.

EXPLORING THE VARIATIONS

Data mining is a "tool" in a journalist's arsenal. Reporters have always used a variety of investigative techniques, from phone calls and beat interviews, to database research and good old intuition to create newsworthy stories. They keep up-to-date on local happenings by routinely contacting local police, hospitals, and emergency managers. When it comes to data mining, adding up the zeros and ones shifts to connecting dots. The process is more than zeros and ones that we plot on flow chart or Venn diagram. Data can be culled to reveal or track our embodied movement and our cultural inclinations in new and different ways. Embedded, organic, fast, visual, clean and scraped data provide new distinctions and multiple stabilities to the facts and figures floating throughout the big data storage clouds and mega hard drives. Once collected and collated through a variety of sophisticated digital processes, this newly extracted and mined data tells a story that we cannot begin to fathom.

Embedded

Data journalism is about finding the story in the data. Embedded data is about finding more of the story or an additional story in data that is not specifically part of the shared facts. Data embedding is one way to increase the quality of information a journalist can use. Every chunk of information carries additional information that, when added together, can create an additional profile of a specific demographic, geographic region, or kind of technology user. Information like how long a user is on a specific website or how many different websites a user has visited, or a users location can be tracked. Sometimes information in the site can be altered and individualized based on the embedded data that is recorded and analyzed on the fly, as the user is online. Information can also be concealed in non-secret text in a process called steganography. This sometimes replaces or supplements encrypted information when culling for data, or brings forward an entirely new story. Access to embedded data can only be made available through sophisticated software processes that reveal the hidden data. A good data journalist will find multiple data sets that help support the consequences of an argument or add validity to a story direction while embracing the unavoidable failings of the data sets themselves. Sometimes the embedded data is the needle in the haystack that breaks open a story. Whistleblowers, people who inform on a group or business doing illicit work, can use embedded data to piece together a direction to pursue to uncover illegal operations.

Organic

Organic data is data that is naturally stored and tracked through the variety of digital transactions that occur throughout the day between humans and devices. As opposed to designed data, the information we have specifically requested and collected through federal tax databases and census reports, organic data is collected through the mouse clicks, search engine, traffic cams, credit card scans and radio frequency identifications (RFID) used everyday in our contemporary digital ecosystem. Journalists do not need to dig for this information because it is naturally created based on our current technology infrastructure. The more we use our connected devices, the more that data is produced and the larger the digital ecosystem becomes. Once a story "breaks" on the web, it has a life of its own that can be tracked through this ecosystem. Sharing, reposting, following, re-sharing, reposting, re-following and rehashing produce a trail of activity. Following the digital path becomes a paramount process for real time updates and breaks in a story. Tracking where the consumer goes for information also becomes part of the process. Information moves easily through different aggregate sites, social news sites, other social media sites or websites. Grabbing a URL and pasting it somewhere, or even easier, clicking a variety of social-sharing buttons facilitate the movement of information. More and more content is being shared every day, increasing the "organic reach" of information. People who pay to have their content pushed out on the web further nudge organic movement of information. In turn, this facilitates the growth of the ecosystem that collects more data about the movement. Organic data, combined with designed data can be a powerful pairing for a digital journalist.

Fast

One of the newer interests in the area of data mining is the ability to get information quickly in to work with and then quickly out to the public. If a story is breaking somewhere in the world, journalists want to know a variety of information about the local environment as quickly as possible to begin searching data. Industry terminology for such fast-moving data tends to be either "streaming data" or "complex event processing." Sometimes information comes too quickly to store everything, so analysis must occur as the data streams in to the server. Sometimes immediate response to an analysis must occur. Open source programs are available to provide batch processing; however, there are also a variety of proprietary products that provide highly customizable options. Quick use of previously compiled and computed information can also be useful for breaking news and investigative stories.

Visual

One of the things that makes data-driven journalism so popular and likable for media use of all kinds is creating visual data that people can better comprehend and understand. Bars, graphs, pie charts and other visual data driven information that can be easily interpreted are increasing in popularity. The visualization turns a math problem or data set into a piece of entertainment. Shiny bright-colored messages and clean, flat-motion graphics or digital shorts abound on every platform from advertising to news to education. These captivating nuggets of information are shared, liked or commented on so data journalists receive a variety of ideas about how interesting the information was for the user. Statistics are hard to understand, and it is much easier when someone else does the heavy lifting and provides the user with an easy way to feel "educated." Diagrams, word trees and maps top the list of data visualization.

Clean

Cleaning data means removing distractions from specific raw datasets. Data cleansing, scrubbing and cleaning is a process of removing incomplete or irrelevant data from a dataset. This incorrect or "dirty" data will make room for either a "cleaner" rendering of the analysis or make room for additional data. Sometimes the data that is collected needs to be altered because it was coded wrong to begin with, and other times the data is specifically wrong or deemed statistically unnecessary. The idea of gathering data is to create factual and usable data to support a media story, and data cleansing provides a more accurate picture of a dataset.

Scraped

Data scraping means moving right from the data to the end user. The process "scrapes" or extracts, data from the digital code so the raw data capture, often in HTML form, can be used. In programming, skills are needed for this process, but data can be easier to access, reorganize and sort when scraped. Removing data from already-available datasets removes the initial collection step from the process. Lists and tables are often not user-friendly when it comes to figuring out meaningful data for a story. A computer program can sometimes reorganize and sort data so trends can be more easily seen. Journalists can look for specific themes and isolate them by scraping them into another Excel spreadsheet to manipulate. Once parameters are set, a variety of programs can sift through data to make it more accessible and usable. The process has been around for years, but new programs have come out in recent years that work to provide even better, more relevant use of data.

The variations for exploring data mining are multiple, and the inroads to discovering a variety of ways to connect the dots using data are continuing to proliferate. This aggregation, collected together with other stories in a mass assemblage, called aggregate news, is discussed in case eleven.

NOTE

1. An early version of this chapter was presented at the Society for the Social Study of Science conference in Copenhagen, Denmark in October 2012.

Chapter Eleven

Case: Aggregate News

Once data is integrated into a story, it is pushed out into the digital world for users. How the content is found, how the story is written and created, and how it is released for consumption is part of our technological texture. Humans are experiencing contemporary information largely in digital form, and if Verbeek explains it rightly, "Humans are what they are, on the basis of the ways in which they can manifest themselves in reality and reality is what it is, on the basis of the way it can be experienced by human beings" (Verbeek 122), then our reality is digital. This case looks at the movement of digital-media content and devices, and how information threads its way through the web to us.

The reality of personal newsgathering is to research news stories through search engines. If a news story does not come up on Google, then it did not exist. Today some outliers compete against the search engine giant, but Google has the lion's share of the market. If an interesting news story does exist, but is in PDF form, it still does not exist for most people because users do not want to download anything. In the age of mobile phones and tablets, downloading is time consuming and eats data. Instead of teaching to the test, today's information is designed to answer the call of the search engine. The data with the best search engine optimization (SEO), the most perfect keywords, and the most sophisticated meta data wins access and exposure. The data without these connections falls to page twenty-seven and is never seen. Users want easy technology and easy content. As Verbeek suggests, users have constructed a reality around their web technology and their lifeworld is experienced through technology. The more technology become embedded in the everyday, the more embedded the pattern becomes and the deeper our technological texture intertwines in the human-technology connection.

The search engine method of vetting material is increasing in popularity. Fewer and fewer people are going to "traditional" paper, radio or television media. Most are going the web. The Pew Internet and the American Life Project's 2014 figures suggest that more than 85 percent of Americans use the Internet. Additional figures suggest that close to 75 percent of developed countries and close to 20 percent of developing counties have Internet access. And new 2015 figures suggest that 65 percent of Americans own a smart phone. There has also been a sizable increase of online news users over the last few years. Users with high-speed Internet connections are more likely to view online news. Younger users are more likely to use online news sources than older users, and those using online news tend to go to non-traditional sites or foreign sites more often. Online users want their news for free and rarely pay for a news service. Last, traditional news sources are still the most popular online sources, but that gap is steadily decreasing.

Online newsreaders self organize to limit the choices of the information available to them through a plethora of individualizing methods. The user chooses to limit his/her news browsing by choosing different sites, choosing a site with predetermined search choices, or by using social media suggestions to tailor choices. The process seems contrary to the reason one might go to an online news site in the first place, to gain information. But maybe not. Maybe people want their own personalized version of the kinds of news they want, but not necessarily lots of news.

Digital-media devices have fostered another kind of newsgathering ability, through mobile phone-optimized websites or phone applications (apps). The news is still aggregated from a variety of news sources, and the notion of news "curation" has been used as a way of explaining how certain news items actually show up on a site. Some news aggregators use algorithms and possibly some human decision-making for headlines and blurb writing, while others focus on interest, choices and relevance to the site's brand. The idea of curating and gathering separate and sometimes disparate elements together is most widely used in cultural heritage organizations like museums, art galleries or library archives. Etymologically, *curating* meant taking care of mortal souls or giving spiritual guidance. The root *cure* means to care for or take care of. The idea of curating digital material like news stories, visuals, web links and other digital assets is relatively new but gaining in popularity. The top news apps clearly highlight what is important to users who want to be informed: the stories are delivered to the user's device for ease of reading/viewing, relevant/personally chosen, and quick updated often. Top choices mesh well with a variety of apps and devices, and most have a quirky side that fosters humor. Some taglines include:

> Feedly: A single place to easily read all the news you rely on to think, learn, and keep ahead.

Fark: Real News. Real Funny.
Pulse: Get up to speed in one newsfeed

These companies are not researching, writing or reporting the news as much as bundling up what is out there in certain ways, and distributing it as a gatekeeper of information, a convenience to users, and a brand or style of newsgathering. And these bundles certainly get used. Increasingly, the disciplines of cognitive science, education and media studies are looking at the ways users engage with news content. Usability studies, site-tracking studies and analytics are all trying to figure out what makes a news site successful. Alain de Botton, in *The News: A User's Manual* (2014), explains it this way:

> After an interval, usually no longer than a night (and often far less; if we're feeling particularly restless, we might only manage ten or fifteen minutes), we interrupt whatever we are doing in order to *check the news*. We put our lives on hold in the expectation of receiving yet another dose of critical information . . . since we last had a look. (I. preface)

LIFEWORLD NEWSROOM

I started following the news as a child, sitting around the dinner table. News was educational for me. And it happened at the same time every evening. I remember when my mother sat me in front of the television and told me to pay attention to the Watergate Trial because "this was history in the making." I was too young to understand what was happening on the screen, but I remembered that TV news content was important and journalists were the authority on information. Later on, I looked forward to a career in news. I went to college for Mass Communication studies and found myself at an internship covering the floods, courtroom dramas and space exploration. After I watching the space shuttle Challenger explode on twenty-two monitors in a television newsroom in 1986, my view of news changed. I saw the variation of the mechanism called news. I noticed that different journalists reported their contently differently. I started to ponder the point of news coverage and the ethics of news agencies. No matter how much or little news there was, it always filled exactly thirty minutes minus commercials. I realized that a subset of decision-makers were in business behind the reporters and anchors and technical crew. I viewed the news with a more critical eye. So I turned to the study of newsmaking. Notes Alan de Botton: "In the developed economies, the news now occupies a position of power at least equal to that formerly enjoyed by the faiths" (2014, 2). He goes on to explain that news has a way of turning the light onto the world's problems without ever turning it back on themselves. From the time we are children, we are asked to share the news in class, watch the news to be more informed and quote the news as a source in school reports. "It is the single most significant force setting the tone of public life and shaping our impressions of the community beyond our own walls" (2014, 3).

My interests in technology and news began to converge in the 1990s when I started surfing the web for news. Since then, in many respects, the national industry has shifted from the "trusted news source" mentality to the "most complete" news source. How might this change the nature of news reading? What does it take to be both informed and complete? This case study works toward opening a space for philosophical reflection on the socio-situated multistability of online news that combines database structure and virtual space/cyberspace exploration. We will explore the human side and the reality as it is present to humans to learn more about the variability of the online news user experience.

INDUSTRY GIANTS

An online user does not need to know the definition of "news" to understand the direction of Google News and what's important to the parent company Google, the owners of YouTube, Blogger, and other smaller entities. The company's mission is to "to organize the world's information and make it universally accessible and useful." The corporate entity says it has no intent to create or own content, but wants only to disseminate it to the public. They algorithmically harvest articles as a gateway to the news, and therefore provide traffic to many, many news sites.

With Google News, launched in beta form in 2002, the goal was simple. The service allows the user to search and browse worldwide news sources, which are updated continuously. Google's tagline was "Comprehensive up-to-date news coverage, aggregated from sources all over the world." The notion was and still is the same, although it has been refined—the Google News search engine user-agent, Googlebot or spider "searches" top hits on news stories around the world and presents them for the user to view based on a variety of ways using the "Crawl and Extraction," "Key words," "press release tag" and "ranking abilities." Google News notes that news publishers can make their site "Google friendly" by following some simple guidelines. There are a variety of ways a user can personalize his or her screen and a variety of ways of viewing the page. The "up to datedness" for each story is displayed below the story headline.

The Google News tagline says, "Aggregated headlines and a search engine of many of the world's news sources." The company's "About Google News" blurb explains:

> Our articles are selected and ranked by computers that evaluate, among other things, how often and on what sites a story appears online. We also rank based

on certain characteristics of news content such as freshness, location, relevance and diversity. As a result, stories are sorted without regard to political viewpoint or ideology and you can choose from a wide variety of perspectives on any given story.

The definition of *aggregate* suggests a total or combined number and etymologically through the Latin *aggregatus*, ad- "to" + gregare "herd," so "to lead to a flock." So Google leads the user to a flock of headlines from the world's news sources. Users are invited to sign in and personalize their Google News front page to "have your news as you like it." You can custom select and deselect news from specific countries and about specific topics and deselect news sources you do not wish to read.

A user can show an increased or decreased number of stories from a variety of news categories like world news, entertainment, sports, health, science and technologies or business and choose local news from a specific region. The personalized section allows the user to rearrange the categories, add a standard or custom section or show headlines only. The user has a choice in the news he or she receives. In short, users can manipulate the site to self-select and self construct the news they are interested in even before the news breaks. Users can even select to follow stories about Google, the company. The Google News personalization pages promote personalized searches based on the users "web history, create and rearrange custom sections for the Google News front page, keep track of the news stories you've read, use Personalized Search to view and manage your history of past searches and news articles you've clicked on. (You can turn off Personalized Search or remove items from your web history at any time) and see news recommended just for you." Beyond setting up personalized news, a user can receive alerts, news to a mobile phone, news feeds and archive searches. The whole idea of getting the latest news shifts the choice from the news room to the user but the media gatekeeping is still present, but shifted to the background through intermediary technology that becomes transparent as it is used.

The news is not made by Google, but it is "Google's News." A traditional television viewer would wait in his or her seat, often as a captive audience, twenty-five minutes to hear the top stories of the day at the evening newscast. Today, users are told they will get the top news when they want it. The mechanism of the Google News search engine entwines a brand and style of usability. Google News offers services like preferences, filtering and language options, number of results, and new results windows because Google News wants "your web search to be exactly the way you want it." The mechanisms of the technology are covered over by the promise of choice, friendly jargon and short instructions to help the user navigate the site.

The early version of Google News claimed to be "A novel approach to news." The word *novel* was originally used to mean new, strange or unusual from Middle French *novel*, but these days Google News has abandoned the novel and instead describe their service as fresh or recent. Google News current version explains it as

> a computer-generated news site that aggregates headlines from news sources worldwide, groups similar stories together and displays them according to each reader's personalized interests. (About Google News).

If a search engine is not neutral, what kind of residue has been left? Google News has come under fire because it seems that their computer-generated sorting does involve human interaction that might taint the "neutral" nature of the site. However, a disclaimer on the site says, "The selection and placement of stories on this page were determined automatically by a computer program. The time or date displayed (including in the Timeline of Articles feature) reflects when an article was added to or updated in Google News." Specifically explained technical guidelines help sites to optimally work to be picked up by a Googlebot. Unique URLs, article formatting and indexing that are optimal by Google's user-agent must be used. Following Google's guidelines for keyword tags or standout tags also helps a news site trying to be user friendly to the Googlebot. A quick search suggests that getting picked up for Google News is not easy and there are many, many rules to follow.

Google News stories are selected based on a numbers game of rank and popularity, but the algorithm also looks for journalistic standards, authority and expertise, accountability and well-written stories. However, several news sources have noted that humans program the algorithm and that items the site management feels are "spam" are not picked up for the site. Starting from the assumption that technology is not neutral, it is important to consider online news sites from a postphenomenological perspective.

BASIC TENETS

One of the main features of online news search engines like Google News is the mediation of the instrument used to pick news stories. Their Google Guide defines Googlebot as a webcrawler that finds and fetches web pages. "Notes Google, 'Crawler' is a generic term for any program (such as a robot or spider) used to automatically discover and scan websites by following links from one webpage to another. Google's main crawler is called Googlebot." An indexer sorts words on pages and a query processor recommends documents it considers to be most relevant. To this end, Google News is "both a microworld and a macroworld which could not be experienced except through the mediation of instruments" (Ihde 2003, 3). Additionally, technology is not neutral; the very act of mediation brings forward bodily, perceptual and cultural changes (13). The user comes to the Google News main page to get started with a search and acts with the variety of links and text, satisficing and scanning, reading and liking. There is a certain asymmetry between the user and the variety of technology that becomes part of the mediated process. Completing a search is an individualistic activity, but many individuals are linked in to the Google culture that holds up the Googlebot. The tool is mediating between the user and the informational world of news texts. And certainly the Google culture of "search" has shaped human activity.

Are digital-media device users shaped by the need to find the best, most up-to-date information to tell the story or the easiest, most effortless way to find an answer? What culture is being shaped? There is perspectivism in this experience and a trajectory of use. One of the fastest growing groups of aggregate news sites are social news websites. The most popular news media sites, like Mashable, Reddit, dig, Slashdot and Buzzfeed, can be shared all over the web at the click of an icon. These news sites act as communities where users submit stories and videos and vote on their favorite news stories, and human editors make choices on featured stories or drop a story that has not received much interest. Users also have the ability to rank user comments about the story, which produces a high level of participation on popular new stories.

Search engines are changing the way we think about finding information, and possibly changing the way specific cultural trends and attitudes are endured and reified (Thorson, 2008). Google and other news search engines like Google Scholar are increasingly upheld as good tools with timesaving elements. Users are taught how to be more sophisticated searchers and to seek strategies for more efficient online searching (Kuster, 2008). It is rare that users are urged to strike out on their own and use the phone or visit someone in person when they can just search for easily digested data online. Search engines are useful and users feel more educated, more media literate and better informed when they "search" for information online. Google News tries not to repeat wire copy stories, but all major news sources and many of the minor ones use wire copy stories because of their money-saving value. News organizations can no longer support a large staff of local reporters, and the converged ownership market that fosters media conglomerates use staff cost effectively across many geographic regions. What might the technological transfer be in this news search engine process? The culling of news to one location seems efficient. The technology functions well. The site is ADA compliant. But as in cases of technological transfer, the non-neutrality is revealed in a variety of ways.

GOOGLEBOT NATION

Aggregated news technology is changing the way we become informed. To be informed in the last millennium, we might have subscribed to several daily newspapers, subscribed to one "big city" newspaper or read it at the library or a bookstore, received several news magazines and viewed local and national news on television or radio. Ten years ago we might have felt that we did our jobs as savvy media consumers if we caught a national or local newscast or two each day and then surfed the web for the newspapers and magazines we trusted. We still received the local newspaper and, possibly, a national paper. Today the Googlebot does it for us, and to be extra informed we go to social sites to become participants in elevating the top news stories to greatness. And the digital media device and its highly personalized content has become the portal.

TAG, YOU'RE IT

In the earlier days of search engine optimization, aggregate news sites used meta-key word tagging. Crawler-based search engines help the website owner more efficiently describe content to the search engines for synthesis. A website owner inserts metatags to the head, or front end, of the site in HTML. Metatags held descriptions and content ratings. The more efficiently you metatagged your site and content, the better possibility it would have been picked up and ranked higher on browsers. Metatags are still used as one of the signals for aggregate news sites, but companies have developed increasingly sophisticated ways to mark news stories to be added to aggregate sites. Algorithms, popularity, optimization and crawlability all effect whether or not web content gets picked up by large aggregate sites. A site owner may create a fine description of the site, but the search engine might change it. Google creates its own descriptions. Some search engines use partial descriptions. Site owners spend a great deal of time perfecting the metatags so their sites get noticed by Googlebot. Only a certain number of sites are listed on the first page of the search engine and each site wants to be one of those few. Creating media sites that are high on the search engine hit counter takes time and money; two things big media have available to them. As a result, most of the big media in America have the most hits on Google News. The Pew Research Journalism Project cites that top news sites include eleven newspapers, six broadcast networks, four news aggregators, three mixed aggregated/original sites and one wire service news site. Google News was ranked in the top ten (The Top 25).

A POSTPHENOMENOLOGICAL TAKE

What is the world-for-humans experience with online news sites? The macroperception lens, a cultural-hermeneutic one, goes far beyond merely "reading" the print or online stories. And the microperception analysis is focused on the sensory/embodiment level. To be sure, the two go hand in hand. Macro and micro dimensionality occur one within the other.

Macroperception

"Reading" Google News is partially based on the screen one views it through and partially based on the experience of experiencing both the web portal and the content being viewed/read/scanned. A macroperceptual focus on aggregate news sites uncovers "the frameworks within which sensory perception becomes meaningful" through interpreted perceptions informed by "the cultural context in which they take place" (Verbeek 2005, 122). When a user looks for news on a digital-media device, context- specific things begin to happen in the lifeworld. Even choosing to "check in" on the news can be related to the device itself. Has the app used a small numeral used a small numeral 5 you that you have missed five news stories? Or is a newsfeed always running on the screen so the user can "check in" right now, in real time, or where he/she left off a few hours ago? As the device mediates the user experience, the perception is one of "a figure against the ground, with every object situated within a surrounding context" (Ihde 1993, 75). The user is in a specific position, most likely face-to-screen, within a contexted lifeworld that places value on being up-to-speed and up-to-date with world happenings. The neighbors say, "Did you see the news today," as if you yourself were featured in a story. Stories are considered "newsworthy" and time is "of the essence" when it comes to reading, reporting and consuming the news. Digital media devices make it "easy" for users to live in this context of "latebreaking" and "time sensitive" event. Their device tells them what and how much they missed while they were living their life. People believe that most news is credible and that they generally know what news is not credible. Personalized news means that users can prioritize content. Individualized content saves time and screens out unwanted news. Sports scores at the top, sports news in the middle and top stories down the right side. Or, world news up top, regional news in the middle, entertainment news at the bottom. Unclick the local news and sports news.

The technological texture of the mediated news environment is individually designed and constantly available. User and the mediated news experience is an intertwining between the user and what is being used and the intent of the use. The user feels culturally in the know, not left out of the loop with his or her device in hand.

Experienced users know the signs of the web world. Underlined or colored text means the headline is hyperlinked to the article. Pull-down windows offer choices. Slide bars allow movement up and down the page. Left bars usually offer more sections, to peruse. A variety of usability studies have allowed sites like Google News to integrate the common language of the web. The site is ADA accessible, and common-source protocols have made many sites similar in look and "feel." But that does not mean that everyone reads or uses Google or Google News the same way. When my son was seven he asked me to search a word for him because he knew that Google would say, "Did you mean?" and would spell the word correctly. So he used Google to check spelling.

Microperception

Aggregate news users are basing their experience from what they know about news media and news content. Through seeing, hearing and touching, the digital media device user can be part of "breaking news." Users feel elated at the recent win of their favorite team, or irritated that their stocks fell or there was an accident on a road they drive regularly. Digital-media device users feel sympathetic about the latest victims of a disaster and angry about recent political changes. The bodily sense of emotions of the device user experiencing news is varied and visceral. As the traditional habits for timed news fades, more options for streaming news become popular. No need to set an appointment with the newscaster to feel privileged with an educated perspective of the world. Aggregate news is "familiarity within the known praxis" of consumers of news media. The body knows and understands the experience, without a "how to guide."

de Botton reflects, "Why do we, the audience, keep checking the news? Dread has a lot to do with it. After even a short period of being cut off from news, our apprehensions have a habit of accumulating. We know how much is liable to go wrong and how fast" (2014, 4). The headlines in the news harken back to the early days of newspapers, and some of today's layouts and apps use the cultural embeddedness of print well. The writers and companies, editors and owners would mostly be the same if the user went to each news source individually. But what about the personal agency to take the time to search the news by individual story? News is no longer a package deal in the same way that singles do not come from albums in the music industry. Each piece of content is its own nugget, with its own trajectory, and the aggregate news organization is the plate from the buffet. Only so much can fit, but it can be piled high on the platter. How might this change the nature of the experience for the user? What is it about news portals and aggregate news sites and apps that make them seem like neutral arbiters of information, a town crier of sorts? And then there is a strangeness of ease. I used to do "all that" but now I can pull up my personalized news page for my top stories, my sports scores, and my business news. I can get the global news and feel the magnitude of a "contemporary pluriculture" (Ihde, 1993). Aggregate news sites truly exemplify this idea through their multiplicity of imagery, cultural and technological fragmentation, fluidity over time and place and bricolage of information (1993, 64) that is "culturally and perspectivally embedded" (65). Who needs the *Wall Street Journal* brought the door each day when I can roll over to my nightstand for my mobile phone to get the news without even getting out of bed? Personally curated news, varied and multiple, at my fingertips. A portal is etymologically understood as "of a gate," brings forward gatekeeping notions in news. Gatekeeping is a kind of filter of stories and keeps back the flow of information, but allows some to filter in through a highly sophisticated algorithm. Many sites share their stance on how they arrive at their choices. Some filter by popularity, some by variety. Filtering means to let out something and allow something else to pass. Gates allow some things in and others out. The Googlebot is the sentinel at the door of the news world, and the goal is to "trick out" your content with the kinds of specifics the bot needs so you are not only in, but also in the top ten to show up on the aggregate news page.

IF IT BLEEDS, IT LEADS

Using aggregated news is an ordinary lifeworld experience for many. Learning about disasters, murders, catastrophes, accidents, sad stories, deceit and other kinds of mayhem and bad news is at the crux of the aggregate news content. As explained by de Botton, "Our background awareness of the possibility of catastrophe explains the small pulse of fear we may register when we angle our phones in the direction of the nearest mast and wait for headlines to appear." (2014, 4). Users surf the web daily without thinking much about it, especially when the aggregate site is their home page, with their email icon in the left corner and their social media at the bottom. Using aggregate news fits within the perceptual phenomenological frame as a "gestalt, or figure against the ground, with every object situated within a surrounding context" (Ihde 1993, 75). As news get "old," time banners click away to let you know when it was originally posted. The content is not static web but a new brand of online experience called "live web." And the savvy media literate user likes receiving three different news sources about the same story to potentially learn multiple sides of an issue. The user experiences primary and secondary qualities in the interaction of the software, clicking links and scanning text, surfing around the sections and clicking more links, within the perceptual lifeworld of home, work, airports, malls, buses, streets and stores—virtually anywhere. But the depth of the perception varies. Am I searching, or browsing, or passing time to look busy?

The world is indeed getting smaller when it comes to digital-media devices. And the device presents some face or facet of itself back to the user. The human-technology relation between a user and a digital media device engaged in newsgathering is a body in mediation.

> Indeed, from the perception of my bodily senses, there is no thing that appears as a completely determinate or finished object. Each thing, each entity that my body sees, presents some face or facet of itself to my gaze while withholding other aspects from view. (Abram 1996, 50)

Technology reflects the way reality is presented to humans in the world, but not in a strictly subject-object experience. Reflects Verbeek,

> By saying that mediation is located "between" humans and world (as in the schema I-technology-world), Ihde seems to put subject and object over against one another, instead of starting from the idea that they mutually constitute each other (2005, 129).

Subject and object are mutually interrelated through the instrument. From a postphenomenological perspective, "The relation between subject and object always already precedes the subject and object themselves, which implies that the subject and the object are mutually constituted in their interrelation" (129–130). There is no user without a device to use. Both the user and the digital media device are brought into the intertwining in specific varied ways and together they acquire a specific shape. Mediation is not a "between" but a mutually new shape. Writes Verbeek, "Mediation shapes the mutual relation in which both subject and object are concretely constituted. Someone who wears eyeglasses, for instance, is not the same without them" (130). This artifactual mediation is a co-shaping. Verbeek states that humans and the world they experience are the *products* of technological mediation, and not the poles strung with devices in between.

In 1999, Marshall McLuhan wrote an essay titled, "The Phonograph: The Toy that Shrank the National Chest." I have always been fond of this particular piece and have used it to think through ideas about the human-technology connection. The essay is largely a historic reflection, and McLuhan has much to say about the non-neutrality of media and technology. In this text I have used the word "user" to explain those who use digital-media devices and content. McLuhan uses the term *electric man* as the nomadic information gatherer, juxtaposed against the Paleolithic man as food-gatherer (111). As chapter 11 comes to a close, with one more case to explore, chapter 12's Self-Tracking, the idea of the electric human becomes all the more meaningful. McLuhan writes:

The telephone: speech without walls.
The phonograph: music hall without walls.
The photograph: museum without walls.
The electric light: space without walls.
The movie, radio and TV: classroom without walls. (111)

And so I wonder,

The digital media: everything without wall?

Chapter Twelve

Case: Self-Tracking

In this final case, the idea of multistability turns toward a very specific context: the use of self-tracking technologies by elite amateur athletes. The intensity of the achievement goals and the technological tools and diagnostic measurement makes this a very focused human–technology connection. The athlete's use of self-tracking technology is not new. Athletes have been measuring length of activity and heartbeats per minute for many years. However, new gear used by elite professionals is now available to amateur triathletes and a new wave of self-trackers is emerging. One of the newer areas of study in the world of human–technology connection is through devices called wearables. New software records workouts, neurotracking and measurement of effectiveness of effort output to maximize total performance during training and race days for both the weekend warriors and elite amateur athletes. Manufacturers partnering with sports physiology and sports medicine help build today's self-tracking technology to alter a triathlete's human physiology, ergonomics, gear biomechanics, sustainable power output and efficiency, aerodynamics, improved bike handling and safety. While some athletes use self-tracking technologies to gain overall fitness, triathletes also use these tools to specifically target and enhance performance and endurance.

Self-tracking, led by The Quantified Self Movement (QS) and the idea that we can learn about our selves through body statistics, has fostered a growing community of data collectors. A variety of self-tracking devices, from Fitbit to the Apple Watch, have hit the mainstream consumer market, and tracking fitness and other body measurements has become a pastime for many. But elite amateur athletes take the technology integration further which is why this case is specifically focused on elite amateur athletes. Computerized bike fits; pedal analysis and gait studies are technologies triathletes use to improve power output and efficiency in competitions to win races for a trophy, medal or bragging rights. New gear and gadgets which mediate athletic performance have provided increased revenue for sports magazines like *Triathlete* and *Runner's World*, which have in turn, proliferated the popularity and interest in multisport.

One thought-provoking point that seems important when thinking about triathlete technology is that triathletes compete at the same exact time, on the same course and with the same conditions as their professional counterparts. This idea of a "mass start" for many triathlete events places pro beside amateur, which is a thrilling opportunity in an athletic world where amateurs are often spectators for the "big game" and are allowed to play the same field on a different day or time. All triathletes in a specific age group start their swim, rack their bikes and chat about their race afterward, standing side-by-side. This changes the nature of the exposure to technology and gear used in races. Amateur athletes have the opportunity to see the gear up close and in person, and then check it out during the vender expo that often occurs at bigger races, the day of the event. All four of the conversants I met with for this study said they paid attention to the pros at their races—who was there, what were they riding, how they did, and what their times were.

This chapter 12 case explores the world of self-tracking technologies that relate to the amateur multisport competition of triathlon.[1] This case focuses on the phenomenal body, the body in reality that engages in human–technology connection to create an ultimate hearing for enhanced performance (Merleau-Ponty 1945, 231). But in the human–technology junction, the body also becomes an object for manipulation, and an arbiter of increased fitness. Elite amateur athletes experience the phenomenal body and lived body, and also measure the objective body for diagnostic information through specific technologies.

> Despite the claim of phenomenologists that the body is inherently a subject, it has never lost its status of an object in competitive sports. And now, another way of understanding the body is being developed—a picture of the body that can be modified indefinitely by technology. (Nagataki 2015, 229)

While elite amateur athletes judge their fitness and readiness from how they feel, they also rely on self-tracking technologies that produce statistics perceived to represent "a true reality in an unquestioning manner" (Nagataki 2015, 232). Body and technology together are reconstituted or co-shaped during training and on race day. As explained by Nagataki in his essay, "How Does Technology Alter Sports?" there are two other aspects to the technological intervention in sports:

1. The boundary between the natural body of an athlete and the tech-aided, cyborg-like body is becoming blurred.

2. The distinction between the real space and the virtual one is disappearing. (2015, 234)

For elite amateur athletes, self-tracking is primarily diagnostic. Triathletes analyze and problem solve training to track ultimate fitness in three different disciplines to achieve the greatest total success for the effort. The experience is tech-aided, yes. And at some levels, based on the diagnostic technology that is used in bike fits and training software programs, the distinction between real and virtual is disintegrating. To be sure, triathletes use their technologies specifically for their non-neutral diagnostic tracking statistics. The object is to alter activity in some way to change a training or racing outcome based on their chosen technology's diagnostic and predictive statistics.

TRI, TRY AGAIN

The USA Triathlon Association website notes twelve reasons why triathlon is gaining in popularity and three play a clear part in this study:

- Increase in resources (websites, books, magazines) that provide assistance/education in getting started.
- Growth in multisport shops and triathlon specific training and racing gear.
- Growth in the number of USA Triathlon certified coaches who are able to provide training plans and individual attention for athletes who need guidance and motivation. (Demographics)

The technology triathletes use is paramount to the way the sport has gained in popularity. But does "use" intersect with self-tracking? Before moving on to the postphenomenological analysis it is important to examine quantified self and self-tracking.

> The quantified self (QS) is any individual engaged in the self-tracking of any kind of biological, physical, behavioral, or environmental information. There is a proactive stance toward obtaining information and acting on it. (Swan, 2013).

The Quantified Self website (quantifiedself.com) features tools like the Fitbit, Digfit and Runkeeper as examples of self-tracking tools. These tools capture the kinds of values and markers that a triathlete uses to track fitness. But I think there is more here than that. Yes, these athletes create self-knowledge through numbers. They engage a variety of technologies to get meaning from their personal data. And in the process, what kind of technological connections unfolds? Recent research continues to probe the human–technology connection of self-tracking to study monitoring practices and bodily analytics and its effects on culture and society (Schull, 2016; Van Den Eede, 2015; Ruckenstein, 2014). For the elite amateur athlete, self-tracking involves three specific disciplines and multiple tracking technologies.

Measuring multisport performance in order to gain fitness and time is a somewhat difficult proposition because the race comprises three disciplines with many variables.

> The triathlon is an example of a successive multiple-event competition where the raw rankings or raw times of athletes, pertaining to each individual component, may not reveal a great deal of information as to which individual discipline plays the most important role in deciding whether a triathlete can win the competition. This is due to the immediate succession of the events: what absolute times or rankings are achieved by the triathletes in each component are not vital. It is the time difference behind the lead athlete in each component that matters. Ranking second by a large time difference in any triathlon component implies a rather small chance for winning the contest compared to ranking fourth or fifth in any discipline with a much smaller time difference. Therefore, preprocessing the times and converting them into differential times becomes crucial to data mining in the triathlon. (Ofoghi et al, 2013)

This small chance of winning creates an environment where athletes seek ways to predict their training effort and the optimal working condition their body needs to be at on race day. A male triathlete can only win a medal if the differential running time is ≤ 26 seconds with approximately 87 percent certainty. For female triathletes, the certainty level is approximately 86 percent and the affordable differential running time for winning a medal is 28 seconds (Ofoghi et al, 2013).

Triathletes cast a wide net when thinking about technologies because technology is found in almost every aspect of the sport, from the mix of the carbon bike material, to the amount of the nutrition in the fuel and hydration liquids, to the apps, GPS watches, cloth design for clothing, wet suits, practice gear, shoes, training diagnostic bike fits and power analyses. When less than thirty seconds makes a huge difference between winning and second place, technology becomes an important element to increase fitness to potentially win. The four amateur athletes I talked with all started slowly moving into the sport but eventually owned a complete collection of gear and gadgets. The four conversants in my study have or use almost all of the following items:

- GPS watch; before that, a basic timer watch
- Heart-rate monitor
- Aero helmet, regular bike helmet
- Wet suit

- Aerodynamic, comfortable training and competition gear
- Bike shoes that clip in pedals—hard carbon sole to facilitate power in pedaling.
- Running shoes
- Goggles
- Water cages or water hydration system in bike
- Bike itself—aluminum, carbon, carbon components, carbon fork
- Training wheels, racing wheels
- Aero bars, which are aerodynamic handle bars
- Saddle—which is the bike seat
- Bike computer
- Power wattage meter
- Trainer or rollers for indoor bike use
- Apps and software, like *Ubersense*
- Online mapping sites
- Online comparison sites, like *Strava*
- Diagnostic fit called a bike fit
- Fuel—water, drinks, goo, gel, power bars

As is the case with nearly all sports-related technologies, data technologies were initially intended as tools for use by the athletic elite, enabling them to analyze their performance in ways never before possible and to push the boundaries of human ability. Endurance runners and cyclists use digital physical activity data collection to optimize physical ability and predictability. "As these technologies become more affordable, they are adopted at all levels of competition. These technologies are now being promoted heavily to sports enthusiasts" (Lee and Drake 2013, 40). In a blog post, the amateur triathlete Gregg Gordon explained it like this: "Unlike the sport of basketball for example, where one can heave a ball toward the rim and perhaps have the good fortune or luck of it going through, there is no such luck to get a person to the finish line, at least in the context of fitness. If you don't prepare, you will not finish; pretty straightforward" (Gordon, 2012). So, triathletes must prepare, or the letters dnf (Did not finish) appear beside their name on the publicly available post-race roster.

Process-wise, a triathlete is timed while swimming, biking and running a pre-defined distance. The event organizers typically record each athlete's time in each discipline, as well as the "transition" times in between them, like the length of time from exiting the water to getting onto the bike and the length of time from getting off the bike to starting the run. Races vary in distances from the short course, or sprint distance, (typically a half-mile swim, 14-mile bike and 5k run); to an Olympic or international distance, which is longer; to a half-Ironman distance, and possibly to the Ironman (2.4-mile swim, 112-mile bike, and 26.2-mile run) (Greg Gordon, 2012). For some, the goal is to complete the endeavor. For others, the idea is to win, to place, to rank nationally for national events or to clock a personal best. Other endurance sports triathletes often compete in include the aqua bike (swim/bike), the duel-athlon (bike/run) and the single sport of running a half marathon or marathon. Elite amateur triathletes illustrate what Tiles and Oberdiek, in their 1995 text *Living in a Technological Culture: Human Tools and Human Values*, explain:

> Our aim should not be to take up a certain stance towards technology, but to see various technologies for what they are, in their varied contexts and without the mystification of supposing them to be either fully under our control or wholly out of control. (1995, 28)

This case brings different nuances to the human–technology connection. What is it like, for elite amateur triathletes to co-shape with technology?

HUMAN–TECHNOLOGY CONNECTION

What is it like to Be with technology in the lifeworld; in an ontological way that feeds one's understanding of all of one's lived world experience? The notion of Being is an important one in phenomenology because it explores the idea of "what it is like to Be." When Heidegger (1962) first raised the question of Being, he was met with objection and curiosity because the concept of Being was considered devoid of real meaning. Being is indefinable in a concrete way, but surfaces as a "seeking for an entity both with regard to the fact that it is and with regard to its Being as it is" (24). This distinction opens the way toward understanding Being-in-the-world with technology. For the triathlete, who is always already caught up in both a technological world and the overall nature of existing within the lifeworld, Being can be complicated. Paulette Robinson, in her dissertation *Within the Matrix: Hermaneutic phenomenological Investigation of Computer Experiences in Web Based Conferencing* (2000), shares: "It is located in the actual. It is not intentional nor the subject in a subject/object split. It is embedded in the socio-cultural context . . . It is the everyday world of activity" (Robinson 2000, 56). Understanding Being, also sometimes explained as existence, is helpful when thinking about Being-with, Being-in, Being-in-the-world and other ways of existing in the human–technology juncture in the world of the technologically in-tuned self-tracking triathlete.

TRACKING THE ATHLETIC SELF

To discuss self-tracking, I transcribed conversations from four amateur elite triathletes who rank in the top 25 percent of regional races or qualified for national championship races through USA Triathlon, the guiding body for triathlons from amateur through professional multisport competition. These athletes are amateurs because they do not compete on the professional circuit or receive money from product endorsements or race wins. All have competed in more than ten triathlons in many of the distances offered. Three have highly specialized day jobs and one has a part-time staff position. Three have completed an Ironman. I have given them alternate names for their recorded and transcribed portions for this reflection.

- Sherrie is 29. She has been competing in triathlons for three years. She started as a runner and moved into triathlons to challenge her fitness. She completed her first Ironman in August 2012 and does several races each year, but not the USATriathlon- ranked races.
- Jim is also 29. He was also a runner, but after standing on the sidelines rooting for his wife, he decided to try a marathon and then a triathlon and "found out he was pretty good at it." He completed his first Ironman in August 2012. He qualified for the USAT long course national distance length race in 2014.
- Nate is 41. He is ranked in the top 27 percent in his age group of all amateur triathletes who have completed at least three USA Triathlete-sanctioned races. In 2012 he competed in seven triathlons and in 2013 he competed in eight bike races and an aqua bike race—which includes swimming and biking but not running.
- Michael is 44. He is ranked in the top 17 percent of all amateur triathletes in his age group, made up of all amateur triathletes who have completed at least three USA Triathlon sanctioned distance races and qualified for the Olympic distance national competition. He completed his first half Ironman in June 2013 and a full Ironman in September 2014.

THREE THEMES

The conversation revealed many human–technology themes but this reflection focuses on the three specific themes of balance, free speed/time and economies of scale for the sport and the technologies. The transcribed narratives illustrate an authentic richness in the human–technology connection, so I have made an effort not to shorten or edit the athletes' contributions. These narratives have not been fully thematized in a hermeneutic-phenomenological methodology, but the themes are clear and resonate with the aim of the case, to illustrate the human–technology connection.

Balance

All four elite amateur athletes discussed finding balance in their lives, in their training and in purchasing their gear. They discussed the input of the energy and time in training versus the output of speed during competition and speed versus endurance versus overall fitness depending on the length of the race and combination of the different disciplines. They talked about how they made decisions to purchase power meters, watches and upgraded bike computers.

> I actually got a power meter over the winter. And the power meter, yes, it is very high tech. I mean, it is a low tech power meter as they go, but a high tech concept in that it is a strain gauge that measures force on the bike or power output on the bike so it can tell you how many watts of power you're putting out at a given time and it calculates a rolling average over a ride and so just like heart rate . . . where speed can be used to set training zones, this power meter will allow you to set training zones and it is supposedly the most accurate because your heart rate can fluctuate based on a lot of different conditions—temperature, hydration, status, fatigue. But power is just purely power output. It's dependent on other variables but it is a directly measurable quantity based on how hard you peddle. (Nate)

> I think I got the Target special Ironman watch, the one she had, for $19.95 or whatever just so I knew the time it took me to run a certain loop. And I just tried to improve upon it from that. Because the loop was constant so I figured that if I worked hard on this section then I can cut down 30 seconds or whatever. So it started from there in running. And then I decided to get the GPS watch when I trained for my first marathon in 2011. But actually I trained for that marathon with just the Ironman watch and I got the GPS watch like two days before the race. So I was like, now I know what pace I'm going to run, before that I would do calculations in my head. If I was running 9 miles at whatever pace, then this was my time in this point in my run, which was kind of a pain but I dealt with it. So the GPS watch pacing wise was perfect for my first marathon. Pretty important I think. (Jim)

> For the road bike, when I got that, it came with a computer but it doesn't have cadence on it. And so I used that one for a little while and it is just basically your speed and your distance. And that's more of a fitness level thing. And then, after a few races, and doing reading, and figuring out what kind of racer I was, I realized that the cadence, how fast you pedal, is my style. I'm not going to be a brute force biker and do all power. It's the endurance. I can pedal fast . . . And so to properly train in that style you need your cadence, your average speed per revolution. And so I got a bike computer that would do my cadence for me and then I train at that specific level. (Michael)

For Sherrie, it was a sliding scale of how much she was "serious" about doing triathlon as a fitness and competition activity.

> I realized, when I was riding my bike, that I didn't know how fast I was going. I'd like to know my mileage per hour. So I bought one of those little bike computers that you mount on your bike and it tells you how far and how fast you're going. And I think I got really serious. Well, so I did that first sprint, and I did the same one the next year and did much better. Then I got shoes and clip in pedals, so I kind of did that little upgrade. So the first two sprints I did I used the same gear, same outfit, and a helmet. So after those two I decided I wanted to go a little bit longer and a few races a year and not the same ones. So the third year I did triathlon I did the Iron Girl down in Columbia, Maryland, and I did Happy Valley up in State College, and Catfish Sprint. Iron Girl was my A race that year. But once we moved up here, I said, let's see if I can go longer and see how it goes. And then I started thinking about the clothes I was wearing. I thought, well if I'm going to go longer, and get serious about this, let's slowly upgrade some stuff. So I think I got some tri shorts instead of your normal Under Armor spandex. I didn't upgrade to the heart rate monitor but I knew I wanted a heart rate monitor because it was a good thing to train with eventually but I didn't do that until I decided to run my first marathon. Because at that point I really wanted to know. (Sherrie)

They also each shared their struggles with ideal training time, technology and race conditions versus real life "putting the time in." Phrases like "getting serious," "putting the time in," "making the race and training easier," and "balancing needs versus wants" figured in the human–technology connection between the elite amateur triathlete and his or her gear.

> But actually when it comes to training I want to be totally mindless about it to be perfectly honest. I mean in a lot of ways, because you know I am thinking so critically in my day job, that I want exercise to be something that's really touchy feely . . . That I really just enjoy . . . to escape from all that. So on some level, although I am very curious about all of the technology, and I have gotten into some of it, I don't use it a lot. When I started running, I did a marathon without anything. I mean I was running next to people and they were beeping, like every five seconds, and I was like, what on Earth is that. And a friend told me, oh, that's their heart rate monitor. They are going above or below their target heart rate. Oh goodness. That's craziness. They were beeping it constantly. But you know it wasn't too long until I had a GPS watch, and I was starting to like, log my miles, my pace. Um, I would actually look at my pace while I was running to try to hold a certain pace. And then it went to getting a heart rate monitor and trying to set heart rate zones . . . based on whether I was doing endurance workout or doing a speed workout. You know so I got into that, and that past three years I was doing triathlons. So I was trying to figure out how to hit certain training zones in swimming that would be different from biking, and running. And it got so complicated that I was starting to trend away from this and you know, the ultimate measure is, you know, how fast can

you go over a certain distance. And so I started to go back to . . . how do I feel? (Nate).

For an elite amateur athlete, decisions shift between "time in training" and "the technology that one can afford." Variables like the conditions during race day and training, upgrades versus obsolescence of technology, training versus racing time, and real opponents one sees on race day all figure prominently into training decisions. Virtual opponents like those seen on websites like Strava, where people log in time on specific segments of a workout and evaluate their time relative to others who have done the same segment, motivate these triathletes to work harder. The times of other competitors at other races are also part of the conditioning and training decisions. Competitions often post times from previous years so searching competitors online to gauge fitness and competitive edge often occurs. Sherrie has found this experience to be interesting:

> There is also one called Strava, so you can go on there and upload your workout and see, basically, if there are segments that people can create, so you can go there to see how you compare with other people who have done the same ride or the same part of the ride depending on how long or short it is. And it's kind of interesting because then once you get to know where the little segments are, it's like, let's see if I can beat them on this one. Them, (laughs) the virtual them, the other people that ride it too. (Sherrie)

Free Speed/Time

A second strong theme is free speed and time. Time can mean time to train, final time at a race or the time that is shaved off through the process of training and technology use. In most cases, free time means the time that a triathlete can shave off of the overall competition time through better gear, gadgets, increased fitness and more specified training.

> I like to be competitive and you start to do pretty well at local events and stuff and you start to realize, oh, I came in 4th place by two seconds but two of the three guys who beat me, beat me by less than a minute and they were racing on three thousand dollar wheel sets and using aero helmets and so you sit down and you do a little research on the computer and realize that just wearing an aero helmet would have saved you 52 seconds on that sized course and you start thinking, well that's free time. Even at my level now I would have come in 2nd. (Michael)

> I looked at number 1, where I could get free speed. So, um, aero, number 1, so I got aero bars for my road bike. Aero helmet. And then I realized, yeah, it does give me a little bit of free speed but I always say and she laughs, it's the Indian not the arrow. So I really worked on my fitness for running and cycling . . . [I] volunteered to be a lab rat for these kids to do a VO2Max test so I

got a free Vo2Max test. I knew where I was at whatever time so that was nice. Having a heart rate analysis done. So I got a heart rate monitor with my first GPS watch and I used my zones to train and improve that. But I don't use it anymore because the battery died again and I'm too lazy to change it. And I've really gone off of feel, how I feel, breathing patterns, read things on Villa news where they talk about a chain lube that could actually get you 8 watts. It's like 8 watts—what's that? I'm like awww, gosh, this is too much. What's an 8-watt difference, you know? And like the bike frame is aerodynamic so they would say that if I got on the bike in a wind tunnel, which is not something I am interested in, and they developed the frame for this. This will see if you get 30 watts in 25 mph. And you go, 30 watts, that's nothing, but you start to realize what 30 watts is and that is a big gain. Like you can train hard all year and get a 20-watt gain in your threshold and be like hey, that's an improvement. So 30 watts is actually pretty big. If it's in real life. And the wind tunnel is such a sterile, if you will, laboratory environment. (Nate)

You can create workouts so you can do an interval workout. So I was like, lets set the interval thing to—so you have your run time so we will set it for a quarter mile and then the rest interval for a quarter mile and we'll repeat that for six times. So then you can set your watch and so all you do is you go run and your watch beeps at you and says you hit a quarter mile this is your rest interval. And then it will count down the seconds until you have to do the next one. So I started using that a lot over the summer too because it kind of broke up the monotony of lets just go run around the track. So I would do that and I would go out on the road and before I knew it four miles was done. I was like wow, that was fantastic, because I was always looking at my watch to look. It also distracted me from thinking about how tired I was or my legs hurt or whatever issue it was that day. It was kind of neat and for Ironman we bought a training book and we went through this 24-week plan so they did a lot of blocks where four weeks you'd have the same kind of run on Thursdays so for that workout if I was doing intervals on the road I could say—maybe I can get a little further down the road if I'm running faster. I wasn't running for pace—just a set distance and time. So eventually I'd get further and further down the road. So that was pretty interesting to see. Let's run faster and see if I can get a little further . . . without really knowing how fast I was going. (Sherrie)

Economies of Scale

Each triathlete I conversed with discussed the tension between the costs of the technology verses their seriousness in the sport and the need to work on personal fitness as well as relying on the technology.

Well part of it is just your desire. You outgrow the technology you have and you want something more. My bike computer was fine at first just doing the miles . . . And then you think, wouldn't it be great if it did this too. Or you read an article where it makes sense and you say, yeah, I see where that would increase my speed or my training level and so you just kind of pick up little bits and pieces or you have a training buddy and then you start to get that feeling of wow, I think that would be cool or that would really help and sometimes it is a help and some times it is just a really cool thing to have and keeping your interest is important. When you are training four to six days a week the new little gadget can keep you motivated to push that next year's training or if you don't get the new thing you start wondering . . . eh, am I still enjoying this . . . so the new piece keeps you going . . . I actually go for training, at least in the races I've done so far, now it might be difference if I'd step up to the Ironman distance, but for the Half [Ironman], Olympic, and Sprint distances I actually go mostly off my body feel versus the technology because except for the half Ironman I did that had some pacing involved in the race, the sprints and the Olympic distance you pretty much just go. And so you just push you just feel with your body and if you're pushing too hard you can back off just a little bit versus using a heart rate monitor or anything like that that would tell you, "Oh, I should speed up a little bit now," or "Oh, I should slow down a little bit now." I use more of the technology in training. I still look at cadence on some of the bike races and on my half Ironman I was actually watching my speed and making sure I wasn't going too fast because I was afraid that it would be bad on my run. So I think that in the longer races I will start to go into the use of technology during the actual race. (Michael)

You know, I've been very interested as I've read about some of the pro triathletes who have done things that have made amazing gains, say, in the Ironman competition in Hawaii. Where one individual, Greg Alexander, swallowed a pill that could measure all kinds of body physiology including temperature. And so they use that to calculate hydration. And they found that if they could keep his core temperature to—because his core temperature actually went up to 102—on a ride they did over there outside of competition—and so they hydrated him up differently based on his core temperature and he was able to maintain higher power numbers, and he was able to go faster and wow that technology is amazing, but not something that's practical for me (Nate).

It gets to a point where you make so many upgrades like, I have a great bike, but what could I do to make this bike better. I could get wheels, new aero bar set to all carbon, I could . . . but I think it also gets to a point where you work on you. Yeah, I can make all of these upgrades to my bike but how much better would that make me. Let's try to work with what I have and try to make myself faster, get better conditioning, get in better shape . . . to structure my workouts to make them more . . . Let's have a purpose with these. Let's do tempo run, let's do speed work, let's do a long run. It's like how can I, me, make myself physically better with training rather than always upgrade and buy the next best thing to make myself better. (Sherrie)

POSTPHENOMENOLOGICAL RENDERING

Elite amateur athletes create self-knowledge through numbers. They co-constitute with a variety of technologies to cultivate personal data to help in the decision making process. The three postphenomenological variations balance, speed and economies of scale reveal experiences that identify technological and instrumentally mediation. For the triathlete, it is a constant and varied, co-shaped connection between technology and self.

- Gauging fitness and endurance in practice through technology.
- Gauging speed, power, distance, cadence and heart rate in competition through technology.
- And then evaluating the performance of all of those values and additionally, the competition conditions, after competition is over.

And training is different from competition. The variables and data points are many and the hours to contemplate are too few because this takes time away from training.

> So I was trying to figure out how to hit certain training zones in swimming that would be different from biking, and running. And it got so complicated that I was starting to trend away from this and you know, the ultimate measure is, you know, how fast can you go over a certain distance. (Nate)

But he also shares this experience during a race,

> I used it for one race but I didn't look at it. Because it was a bike race and things are changing dynamically, and so sometimes the group is going slower and your power numbers drop. And sometimes the group is going very hard and you are trying to stay with the group so I would never say, well, they're doing 450 watts and that's out of my zone because as soon as I do that I drop and then I am done and lost. (Nate)

And he goes on:

> So say I've got an hour and I wanna do intervals on the bike, especially in the winter, then I will use the power meter and say—I wanna do Vo2 max intervals. I know what my Vo2 max power range is and I'm going to try to hit those numbers for five minute intervals for example. And that's been neat. That's what I really like about that tool. It keeps me focused. It keeps me interested and engaged in a work out. And then over time I can reassess those areas those ranges. . Have I had progress, have I stayed in the same spot? So I hope to use it that way, but many people use them in a race. For example, I'm doing a triathlon, and I know my threshold power and I wanna stay right at that because I know I can run this fast after two and a half hours at that power output. You know, so they can use that data during a race to help them calculate or kind of anticipate the next leg of the race so that would be ideal in a triathlon. (Nate)

The micro-perceptual experience of being a triathlete can be an experience of the sensory body saying, you're going too strong because you can hardly breath, juxtaposed on the heart rate monitor that is beeping when the triathlete goes out of the target heart rate zone, layered with the macro-perceptual hermeneutic reading of the GPS watch showing the time in competition while the triathlete performs the cultural experience of being an elite amateur athlete, while interpreting the hermeneutic rendering of the bike power meter which is showing a higher wattage than the athlete can sustain. And the triathlete is taking all of this in as he or she completes three disciplines in a row with short transitions for gear changes. No wonder that it is a cultural ritual that triathletes who complete an Ironman come off the course and go to the tent to get an Ironman tattoo to mark their achievement in a very corporeal way. The variational interplay reveals additional structural multistabilities that bring forward increased insights about the triathlete's experience with technology. It makes sense that so many of the triathlon magazines promote gear that makes technology increasingly easier to use. The GPS Garmen 910 gives the triathlete statistics on the swim, bike and run. The GPS Forerunner watch beeps to vary workout so the athlete does not need to consistently read the watch. The intellectual fatigue of analyzing date while competing, which Michael calls "doing the math," can increase the time it takes a triathlete to complete the course. The triathlete's self-tracking is not completed while sleeping or completing daily chores, but two-thirds of the data is gathered during either training or competition so the variables for when the data is gathered are different, but the experience of training is supposed to infer increased fitness and endurance, i.e. speed and results during competition. The ways of reading the data and the experience are multiple and varied and perceptually based, while the data and experience themselves are multiple and varied. In *Postphenomenology: Essays in the Postmodern Context*, (1993), Ihde explains that "To both 'see' in an embodied position, and to 'read' in an apparent position, and to be able to easily 'hermeneutically' transpose between the two positions is part of what it means to perceive in the now *postmodern* lifeworld . . . *both* deconstructive *and* yet structural" (87). The elite amateur co-shapes with many technologies that measure different things and collates them to gain a holistic digital picture of fitness and endurance over a variety of variables. Their experiences illustrate new insights about the human–technology intersection, provide examples for technological mediation at its broadest sweep and is a good case study for exploring peak performance in our doubled, co-shaped, digital media: human–technology connection.

NOTE

1. An early version of this chapter was presented at the Society for the Social Study of Science Conference in San Diego, California, in October 2013.

Epilogue

Convergence: Revising the Texture

A tapestry is a textile art of interwoven threads that come together to form a complete picture or pattern. Early etymological links to the word *tapestry* mean a heavy covering. I wonder, is the technological weave in our contemporary world a heavy covering? Are we irrevocably in the midst of our technologies, and how might the entanglement affect our future lifeworld experience? And what of the texture that has been revealed? Its quality and pattern is regularly appearing in all facets of the lifeworld and humans have to purposefully work to untangle from it. We may see the technological texture before us as a tapestry: pretty, or rich, or interesting, or intricate, or heavy, or light. But it is difficult to see through it. Some untangle by unplugging while others use stronger phrases like digital detox or disconnecting to reconnect. The technological texture is so regularly patterned within the human–technology connection as a given instead of a choice, that it has become invisible or backgrounded.

This final chapter is more about continuing the conversation than providing a conclusion, because postphenomenology does not advance conclusions, but explores trajectories, variations and multistabilities. The methodological underpinning does not analyze technology in order to seal the fate of the topic but probes to open the weave to unseen possibilities. So in effect, this epilogue is an opportunity. There is much work still to be done in this ever-changing world of digital media.

One growing area of interest for research and development of digital media technology is called *responsible innovation*. In their chapter "The Emerging Concept of Responsible Innovation: Three Reasons Why It Is Questionable and Calls for a Radical Transformation of the Concept of Innovation," in *Responsible Innovation 2: Concepts, Approaches, and Applications* (2015), Vincent Blok and Pieter Lemmens explore the concept of responsible innovation and work through whether or not responsible innovation is even possible. Their conclusion is worth pause.

> The main difficulty of responsible innovation revolves around the response-ability of actors in the innovation process, due to 'epistemic' factors like the inherent complexity, uncertainty and unpredictability of technological innovation on the one hand, and 'moral' and 'political' factors like conflicting worldviews, interests and value systems among stakeholders and power imbalances on the other. We concluded that the practical applicability of the concept of responsible innovation is highly questionable (Blok and Lemmens 2015, 31).

Their conclusion has left me wondering. Is the digital media environment, with its inherent complexity, Wild West innovation fueled by crowd-funding campaigns and angel investors and conflicting governments and worldviews able to responsibly innovate? And why might it be important to consider this idea in light of the global social, political and ethical strands that weave into the technological texture?

This book is an exercise in examining digital media's human–technology connection to recognize that digital media, its technology and its content do not lack effect. Responsible innovation, however "pie in the sky" it may seem, is an important consideration for the future of digital media innovation. How long can we allow a non neutral technology to knit within the fabric of our very being, before we examine it more closely and seriously? Digital media is not invisible or neutral. It textures our world in multiple and varied ways. This acknowledgement is an important one. A technologically textured world is not broken down into positive and negatives, but studying it's design is an important exercise in identifying micropercptual (embodied) and macropercpetual (socio-cultural) themes to recognize them for the power they have. The case studies in *Digital Media: Human–Technology Connection* worked to examine a tapestry that co-constitutes and co-shapes the entanglements that illustrate a transformative difference. The "Raw Materials" section (chapters 1, 2 and 3) introduced the metaphor of technological texture, explored the Philosophy of Technology and introduced the philosophical terms: perception, embodiment, lifeworld and interrelational ontology. The holistic definition of digital media was examined, and phenomenology and postphenomenology were explained. Section 2 "Feeling the Weave" used the postphenomenological lens to examine nine different digital media case studies. Chapter 4 case, "The Screen," explored the frame-screen-window metaphor of the human–technology connection. Chapters 5, 6 and 7 explored digital sound. The chapter 5 case, "Dwelling in Digital Sound," explored the variations of filtered sound and real/pure sound, chapter 6 case, "Earbud Embodiment," explored the phenomenological multistabilities of earbuds as wired for music listening, noise canceling, as accessory, and avoidance of outside noise and interpersonal communication. Chapter 7's case, "Portable Sound" continued the idea of wearable sound and explored the question "Why and how is who communicating what to whom and with what effect?" The final section studying sound, the chapter 8 case, "Dubstep," explored the variations of language, performance, and collaboration. Chapter 9's case, "The Photo Manipulation Aesthetic," explored four different variations of photo manipulation; chapter 10's case, "Data Mining," explored ways data alters the information journalists use and the process by which the statistics are gathered. Chapter 11's case, "Aggregate News," continued the idea of newsgathering through a macro-perception lens for "reading" the print or online stories and the micro-perception analysis focused on the sensory/embodiment level. The final case, chapter 12's study of self-tracking, analyzed wearable technologies used by elite amateur athletes. In all of these ways technologies change the human–technology-world experience. They mediate practice, shape human experience, and alter use and lifeworld perception.

In our world, this tapestry of technologies is called *convergence*, the merging of many distinct technologies, devices and industries. The world is both converged and converging because there is always a new technology to integrate into the weave of life. Understanding our place in this tapestry is the important part. It requires thinking about, understanding and adjusting or not adjusting in our daily lives. Digital media is not neutral. We are co-shaped by our technologically world. We are the threads and the background, the weft and the warp. *Digital Media: Human–Technology Connection* has been an exploration of case studies that highlight the technological texture to examine the weave.

I have learned to feel the texture of technology to begin anew to understand digital media in different non-neutral ways and to continue to pay attention and not become complacent in the technological texture. I invite you to do the same. We feel the inexplicable urge to reach out and feel the texture of technology, whether we realize it or not. We create, communicate, collaborate and coexist with digital media every day, and for some, every minute of every day. We work and play with digital media. Shouldn't we be paying more attention to the variations in our human–technology connection because of the technological texture? Technology can be enveloped and explored but not negated or ignored. An empowered stance recognizes our way of being-in-the-world and allows expression within the fabric of the texture. This creates elasticity within the environment that asks questions and lives in the weave for answers. This fabric becomes a new imprint, a new pattern and a new shade across the interwoven strands of creative work as the texture makes a new fold.

References

Aarseth, E. (2003)."We All Want to Change the World: The Ideology of Innovation." in *Digital Media. Digital Media Revisited: Theoretical and Conceptual Innovation in Digital Domains.* Boston: MIT Press. 318–321.

"About Google News." *About Google News.* Google, n.d. Web. 1 June 2015. http://news.google.com.au/intl/en_au/about_google_news.html.

Abram, D. (1996). *The Spell of the Sensuous.* New York: Pantheon Books.

Adatto. K. (2008). *New Image Consciousness in American Culture.* Princeton, NJ: Princeton University Press.

Agliata, D., and S. Tantleff-Dunn. (2004). "The Impact of Media Exposure on Males." *Journal of Social and Clinical Psychology.* 23.1. 7–22.

American Medical Association. *AMA Adopts New Policies at Annual Meeting. American Medical Association.* AMA, 21 June 2011. Web. 2 Apr. 2014. https://www.ama-assn.org/ama/pub/news/news/a11-new-policies.page.

Anderson, C. A., Shibuya, A., Ihori, N., Swing, E. L., Bushman, B. J., Sakamoto, A., et al. (2010). Violent video game effects on aggression, empathy, and prosocial behavior in Eastern and Western countries. Psychological Bulletin, 136, 151–173. doi:10.1037/a0018251

Annenberg Media Exposure Research Group. (2008). Linking measures of media exposure to sexual cognitions and behaviors: A review. Communication Methods and Measures, 2, 23–42. doi:10.1080/19312450802063180

Anton, C. (2001). *Selfhood and Authenticity.* Albany: SUNY Press.

Attias, B. (2011). "Subjectivity in the Grove: Phonography, Digitality, and Fidelity." *DJ Culture in the Mix: Power, Technology & Social Change in Electronic Dance Music.* New York: Bloomsbury. 15–50.

Balsamo, A. (2002). "On The Cutting Edge." *The Visual Culture Reader.* London: Routledge. 685–695.

Beauty Standards around the World. Dir. Esther Honig. *YouTube.* YouTube, 6 June 2014. Web. 18 Jan. 2015. https://www.youtube.com/watch?v=RT9FmDBrewA.

Barnhart, R. (2001). *Chambers Dictionary of Etymology.* New York: H.W. Wilson Company.

Bartneck, C., Hoek, M. v. d., Mubin, O., and Mahmud, A. A. (2007). *"Daisy, Daisy, Give me your answer do!"—Switching off a robot.* Proceedings of the 2nd ACM/IEEE International Conference on Human-Robot Interaction, Washington DC. 217–222.

Beck, Ulrich. (1998). "Politics of Risk Society." *Debating the Earth: The Environmental Politics Reader.* Oxford: Oxford University Press. 587–595.

Berendt, J. (1992). *The Third Ear: On Listening to the World.* New York: Henry Holt.

Blok, V., and Lemmens, P. "The Emerging Concept of Responsible Innovation. Three Reasons Why It Is Questionable and Calls for a Radical Transformation of the Concept of Innovation." *Responsible Innovation 2*. Place of Publication Not Identified: Springer, 2015. 19–35.

Bolter, J. D. & Grusin, R. (1999). *Remedition: Understanding New Media*. Boston: MIT Press.

———. (1996). "Remediation." *Configurations*, 4(3), 311–358. http://dss-edit.com/plu/Bolter-Grusin_Remediation.pdf

Borgmann, A. (1987). *Technology and the Character of Contemporary Life*. Chicago: University of Chicago Press.

Bostrom, N. (2005). "A History of Transhumanist Thought." *Journal of Evolution and Technology*. 14.1. 1–25.

Bradbury, R. (1967). *Fahrenheit 451*. New York: Simon and Schuster, 1967.

Brey, P. (1998). "Space-shaping technologies and the geographical disembodying of place." *Philosophies of Place*. Lanham, MD: Rowan & Littlefield. 239–263.

Brown, J. D., L'Engle, K. L., Pardun, C. J., Guo, G., Kenneavy, K., and C. Jackson. (2006). Sexy media matter: Exposure to sexual content in music, movies, television, and magazines predicts black and white adolescents' sexual behavior. Pediatrics, 117, 1018–1027. doi:10.1542/peds. 2005. 1406

Brugioni, D. A. (1999). *Photo Fakery: The History and Techniques of Photographic Deception and Manipulation*. Dulles, VA: Brasseys.

Buckingham, David. (2008). Youth, Identity and Digital Media. Boston: MIT Press.

Bull M. (2008). *Sound Moves: iPod Culture and Urban Experience*. New York: Routledge.

———. (2006). "Investigating the Culture of Mobile Listening: From Walkman to IPod." Consuming Music Together: Social and Collaborative Aspects of Music Consumption Technologies. 35. 131–149.

———. (2004). "Automobility and the Power of Sound." *Theory, Culture & Society*. 21. 243–259.

——— (2000). *Sounding Out the City*. New York: Berg.

Cage, J. (1979). "The Future of Music." *Philosophy of Media Sounds*. New York: Atropos Press. 113–130.

Casey, E. (1993). *Getting Back into Place: Toward a Renewed Understanding of the Place-World*. Bloomington: Indiana University Press.

Casey, E. (1997). *The Fate of Place: A Philosophical History*. Berkeley: University of California Press.

Chaieb, L., Wilpert, E. C., Reber T. P., and J. Fell. (2015). Auditory beat stimulation and its effects on cognition and mood States. Front Psychiatry. 2015 May 12;6:70. doi: 10.3389/fpsyt.2015.00070. eCollection 2015.

Coleman, E. G. 2010. Ethnographic Approaches to Digital Media. Annual Review of Anthropology. June 21, 2010. http://www.annualreviews.org.proxy-millersville.klnpa.org/doi/pdf/10.1146/annurev.anthro.012809.104945

Couch, L. W. (2012) *Digital and Analog Communication Systems*. (8th ed). Upper Saddle River, NJ: Prentice Hall, 2001.

Couldry, N. (2008). "Mediatization or Mediation: Alternative Understandings of the Emergent Space of Digital Storytelling," *New Media & Society* 10.3. 373–391.

Crease, R. P. (2006). "From Workbench to Cyberstage." *Postphenomenology: A Critical Companion to Ihde*. Albany: SUNY Press, 221–230.

Cubitt, S. (2009). "Case Study: Digital Aesthetics." *Digital Culture: Understanding New Media*. New York: McGraw-Hill. 23–29.

Culkin, J. (March 18, 1967). "A Schoolman's Guide to Marshall McLuhan." *Saturday Review*, 51–53.

Creeber, G. and M. Royston (2009). *Digital Culture: Understanding New Media*. New York: McGraw Hill.

Csikszentmihalyi, M. (1990). *Flow: The psychology of optimal experience*. New York: Harper-Perrenial.

Data Science and Big Data Analytics: Discovering, Analyzing, Visualizing and Presenting Data. (2015).S.l.: Wiley.

References

Davidson, D. (1981). "What Metaphors Mean." In *Philosophical Perspectives on Metaphor*. Ed. M. Johnson. Minneapolis: University of Minnesota Press. 200–220.

de Botton, A. *The News: A User's Manual*. Vintage, 2014.

Dekker, A. (2003). "Synaesthetic Performance In The Club Scene." *Cosign 2003: Computational Semiotics Proceedings.* Cosign Conferences. United Kingdom: Teesside University.

USA Triathlon. "Demographics." Accessed August 29, 2015.

Deuze. Mark. 2006. "Participation, Remediation, Bricolage: Considering Principal Components of a Digital Culture." *The Information Society* 22 (2), 63–75.

Dewdney, A. and P. Ride. (2014). *The Digital Media Handbook*. New York: Routledge.

Donath, J. (2000). "Being Real: Questions of Tele-Identity*.*" *The Robot in the Garden* Boston: MIT Press. 296–311.

Dorfman, E. (2009). "History of the Lifeworld." *Philosophy Today*. 53 (3) 294–303.

Dusek, V. (2006). *Philosophy of Technology: An Introduction*. Malden, MA: Blackwell.

Dyson, F. (2009). Sounding New Media: Immersion and Embodiment in the Arts and Culture. Oakland: University of California Press.

Eisenstein, S. (1969). *Film Form: Essays in Film Theory*. Translated by Jay Leyda. Evanston, IL: Northwestern University Press.

Evens, A. (2012). Web 2.0 and the Ontology of the digital. http://www.digitalhumanities.org/dhq/vol/6/2/000120/000120.html

Friedberg A. (2006). *The Virtual Window: From Alberti to Microsoft*. Boston: MIT Press.

Foltyn, J. (2011). "Corpse Chic: Dead Models and Living Corpses in Fashion Photography," *Critical Issues: Fashion Forward*, Oxford: Inter-Disciplinary Press.

Forss, A. (2012). "Cells and the (Imaginary) Patient: The Multistable Practitioner-Technology-Cell Interface in the Cytology Laboratory," Medical Health Care & Philosophy. New York: Springer Science & Business Media.

Friesen, N. and T. Hugg. (2009). The Mediatic Turn: Exploring Consequences for Media Pedagogy. In K. Lundby (Ed.). *Mediatization: Concept, Changes, Consequences*. New York: Peter Lang. 64–81.

Friis, J. K., B. O. and R. P. Crease. (2015). *Technoscience and Postphenomenology: The Manhattan Papers.* Lanham, MD: Lexington Books.

Gadamer, H. (2000). *Truth and Method*. Translated by J. Weinsheimer & D. Marshall. New York: Continuum.

Gallagher, S. (2012). *Phenomenology*. New York: Palgrave Macmillan.

Gant, S. (2011). *We're All Journalists Now: The Transformation of the Press and Reshaping of the Law in the Internet Age.* Washington, DC: Free Press.

Gergen, K. (1991). *The Saturated Self: Dilemmas of Identity in Contemporary Life*. New York: Basic Books.

Ghosh, M., and R. Ghosh. (2007). *Language and Interpretation: Hermeneutics from East-West Perspective*. New Delhi: Northern Book Centre.

Gladwell, M. (2000). *The Tipping Point: How Little Things Can Make a Big Difference*. Boston: Little, Brown.

Goeminne, G., and E. Paredis. (2011). "Opening up the In-between: Ihde's Postphenomenology and Beyond." *Foundations of Science Found Sci* 16.2-3: 101–107.

Gordon, G., (2012)."Why I do Triathlons." http://magnacare.typepad.com/blog/2012/02/why-i-do-triathlons.html. Retrieved 5/6/13.

Gunn, J. and M. Hall. (2008). "Stick It In Your Ear: The Psychodynamics of iPod Enjoyment," *Communication and Critical/Cultural Studies*, Guptill Publication. 5. 2. 135–157.

Habermas, J. (1989). *The Theory of Communicative Action*. Oxford: Polity.

Hansen, M. B. N. (2006). *New Philosophy for New Media*. Boston: MIT Press.

Harrison, K. and V. Hefner. (2014). "Virtually Perfect: Image Retouching and Adolescent Body Image." In *Media Psychology*. 17.2. New York: Routledge.

Haraway, D. (1991). *Simians, Cyborgs & Women*. New York: Taylor & Francis.

Hasse, C. (2008). "Postphenomenology: Learning Cultural Perception in Science." Human Studies. New York: Springer.

Hawkins, N., Richards, P. S., Granley, H. M. and D. M. Stein. (2004). "The Impact of Exposure to the Thin-Ideal Media Image on Women." *Eating Disorders* 12.1. 35–50.

References

Heidegger, Martin, (1977). *The Question Concerning Technology and Other Essays.* Translated by W. Lovitt. New York: Harper Torch books.

Heidegger, Martin. (1962). *Being and Time.* Translated by John Macquarrie & Edward Robinson. New York: Harpers and Row Publishers.

Heidegger, Martin .(1953). *Being and Time.* Translated by Joan Stambaugh. New York: Harpers.

Herrera, M. (2011). "Skrillex Isn't Surprised By Dubstep Takeoff." *Rolling Stone Magazine.* November 2, 2011.

Hickman, L. A. (2008). "Postphenomenology and Pragmatism." *Techné: Research in Philosophy and Technology.* 12.2. 99–104.

Holmstrom, A. J. (2004). "The Effects of the Media on Body Image: A Meta-Analysis." *Journal of Broadcasting & Electronic Media.* 48.2. 196–217.

Husserl, E., and L. Hardy. (1999). *The Idea of Phenomenology: A Translation of Die Idee Der Phänomenologie, Husserliana II.* Dordrecht, The Netherlands: Kluwer Academic

Husserl, E. (1980).*Collected Works.* The Hague: M. Nijhoff.

Husserl, E. (1965). *Phenomenology and the Crisis of Philosophy.* Translated by Q. Lauer. New York: Harper Torchbooks.

Holmstrom, A. (2004). The Effects of Media o Body Image. *Journal of Broadcasting and Electronic Media.* 48.2. 196–217.

Horowitz, Seth. 2013. *The Universal Sense: How Hearing Shapes the Mind.* New York: Bloomsbury USA.

Ihde, D. (2009). Postphenomenology and Technoscience: The Peking University Lectures.

———. (2008). "Introduction: Postphenomenological Research." *Human Studies* 31.1: 1–9.

———. (2007). *Listening and Voice: Phenomenologies of Sound.* 2nd ed. Albany: SUNY Press.

———. (2006). "Forty Years in the Wilderness." *Postphenomenology: A Critical Companion to Ihde.* Ed. Evan Selinger. Albany: SUNY Press, 267–290.

———. (2002). *Bodies in Technology.* Minneapolis: University of Minnesota Press.

———. (2000). *Nature* 406.6791, 21. Web.

———. (1999). *Expanding Hermeneutics: Visualism in Science.* Evanston, IL: Northwestern University Press.

———. (1986). *Experimental Phenomenology.* Albany: SUNY Press.

———. (1995). *Postphenomenology: Essays in the Postmodern Context.* Evanston, IL: Northwestern University Press, 1995.

———. (1990). *Technology and the Lifeworld: From Garden to Earth.* Bloomington: Indiana University Press.

———. (1983). *Existential Technics.* Albany: SUNY Press.

———. (1978). *Technics and Praxis: A Philosophy of Technology.* Netherlands: Springer.

Ihde, D. and E. Selinger. (2003). *Chasing Technoscience: Matrix for Modernity.* Indianapolis: Indiana University Press.

Introna, L. D. and Ilharco, F. M. (2004). "The Ontological Screening of Contemporary Life: A Phenomenological Analysis of Screens." *European Journal of Information Systems Eur J Inf Syst, 13* (3), 221–234.

Irwin, S. (2015). "The Dubstep Mash-Up." *Technoscience and Postphenomenology: The Manhattan Papers.* Lanham, MD: Lexington Books.

———. (2014a) "Technological Reciprocity with a Cell Phone." *Techné: Research in Philosophy and Technology.* 18.1. 10–19.

———. (2014b)."Embodied Being: Examining Tool Use in Digital Storytelling." *Tamara: Journal for Critical Organization Inquiry.* 12.2. 39–49. Web.

———. (2010). "Technological Texture: A Phenomenological Look at the Experience of Editing Visual Media on a Computer." *Phenomenology and Media.* 221–232. Web.

———. (2010b). "The Poetics of Blurred Boundaries." *Proceedings of the Media Ecology Association.* Media Ecology Association, 2010b Web. 21 Jan. 2014. http://www.academia.edu/684884/The_Poetics_of_Blurred_Boundaries.

Irwin, S. (2005). "Technological Other/Quasi Other." *Human Studies.* 28.4. 253–267. New York: Springer.

Jenkins, H., S. Ford, and J. Green. (2013). *Spreadable Media: Creating Value and Meaning in a Networked Culture*. New York: New York University Press.

Johnson-Laird, P. (1988). *The Computer and the Mind*. Cambridge, MA: Harvard University Press.

Jordan. P. (2002). *Designing Pleasurable Products: An Introduction to the New Human Factors*. Philadelphia, PA: Taylor & Francis.

Juslin, P. N., and J. Sloboda. (2010). *Handbook of Music and Emotion: Theory, Research, Applications*. Oxford: Oxford University Press.

Kaplan, D. M. (2010). *Readings in the Philosophy of Technology*. Lanham, MD: Rowman & Littlefield.

Kember, S. and J. Zylinska. (2012). *Life After New Media: Mediation as a Vital Process*. Cambridge, MA: MIT Press.

Kaptelinin, V., and B. Nardi. (2006). *Acting with Technology: Activity Theory and Interaction Design*. Cambridge, MA: MIT Press.

Kovarkik. B. (2011). *Revolutions in Communication: Media History from Gutenberg to the Digital Age*. New York: Bloomsbury Academics.

Krahé, B., Moeller, I., Kirwil, L., Huesmann, L. R., Felber, J., and A. Berger. (2011). Desensitization to media violence: Links with habitual media violence exposure, aggressive cognitions, and aggressive behavior. Journal of Personality and Social Psychology, 100, 630–646. doi:10.1037/a0021711.

Kurzweil, Ray. (2005). *The Singularity Is Near: When Humans Transcend Biology*. New York: Viking.

Kuster, J. (2008). "Become a Sophisticated Searcher." *ASHA Leader*. 13. 9. 44–45.

Lakoff, G. and M. Johnson. (1999). *Philosophy of the Flesh*. New York: Perseus Books Group.

———. (1981). "Conceptual Metaphor in Everyday Language." In M. Johnson (Ed.), *Philosophical Perspectives on Metaphor*. 286-325. Minneapolis: University of Minnesota Press

———. (1980). *Metaphors We Live By*. Chicago: University of Chicago Press.

Langsdorf, L. (2015). "Why Postphenomenology needs a Metaphysics." *Postphenomenological Investigations: Essays on Human-Technology Relations*. Lanham, MD: Lexington Books.

Lavallee C. F., Koren S. A., and M. A. Persinger. (2011). "A Quantitative Electroencephalographic Study of Meditation and Binaural Beat Entrainment." *Journal of Alternative and Complementary Medicine* (New York, N.Y.), 17 (4), 351–355.

Lawson. C. (2008). Technology and the Extension of Human Capabilities. *Journal for the Theory of Social Behavior*. Vol. 40 (2). 207–223

Leder, D. (1990). *The Absent Body*. Chicago: University of Chicago Press.

Lee, V. R., and J. Drake. (2013). "Digital Physical Activity Data Collection and Use by Endurance Runners and Distance Cyclists." *Technology, Knowledge and Learning Tech Know Learn* 18.1-2 (2013): 39–63.

Levin, David. (1989). *The Listening Self: Personal Growth, Social change and the Closure of Metaphysics*. New York: Routledge.

———. (1985). *The Body's Recollection of Being*. Boston: Routledge & London: Martinus Nijhoff Publishers.

Liberati, N. (2015). "Technology, Phenomenology and the Everyday World: A Phenomenological Analysis on How Technologies Mold Our World." *Human Studies*. New York: Springer.

Liberati, N. (2012). "Improving the Embodiment Relations by Means of Phenomenological Analysis on the 'Reality' of ARs." *ISMAR-AMH*, 2012, 2013 IEEE International Symposium on Mixed and Augmented Reality—Arts, Media, and Humanities (ISMAR-AMH), 2013 IEEE International Symposium on Mixed and Augmented Reality—Arts, Media, and Humanities (ISMAR-AMH) 2012,13-17, doi:10.1109/ISMAR-AMH.2012.6483983

Liestol, G. (2003). *Digital Media Revisited: Theoretical and Conceptual Innovations in Digital Domains*. Cambridge, MA: MIT Press.

Magee, H. (2012). The Ethics of Digital Photo Manipulation: Alterations in Pursuit of "Beauty." Brandeis University. Retrieved from http://www.brandeis.edu/ethics/ethicalinquiry/2012/August.html

McLuhan, M. (1964). *Understanding Media: The Extensions of Man.* New York: Signet Books.

McLuhan, M., Hutchon, K., and E. McLuhan. (1980). *Media, Message, and Language.* Skokie, IL: National Textbook Company.

Manovich, L. (2002). *The Language of New Media.* Cambridge, MA: MIT Press.

Mazis, G. (2002). *Earthbodies: Rediscovering our Planetary Senses.* New York: SUNY Press.

———. (2008). *Humans, Animals, Machines: Blurring Boundaries.* New York: SUNY Press.

Merleau-Ponty, M. (2000). *Phenomenology of Perception* (Translated by C Smith.). Great Britain: Routledge.

———. (1987). *The Visible and the Invisible.* Translated by A. Lingis. Evanston, IL: Northwestern University press.

———. (1964). *Signs.* Evanston, IL: Northwestern University Press.

Moran, D. (2000). *Introduction to Phenomenology.* New York: HarperCollins.

Munster, A. (2011). *Materializing New Media.* Lebanon, NH: Dartmouth College Press.

Nagataki, S. (2015). "How Does Technology Alter Sports?: Body, Space and Ethics." *Technoscience and Postphenomenology: The Manhattan Papers.* Lanham, MD: Lexington Books. 211–224.

Nagel, C. (2010). Empathy, Mediation. Media. Phenomenology and Media: An Anthology of Essays from Glimpse: Society for Phenomenology and Media.

Neff, J. "National Ad Division Goes After Retouching of Beauty Ads." *Advertising Age News RSS.* Crain Communications, 15 Dec. 2011. Web. 28 Jan. 2012. http://adage.com/article/news/national-ad-division-retouching-beauty-ads/231620/.

Nóbrega, R., Moura, D., Calcada, T., Trigueiros, P., and A. Coelho. (2014). "Interacting With The Augmented City: Sensors and Mixed Reality." NordiCHI'14 Workshop: Making Places: Visualization, Interaction and Experience in Urban Space. October 26, 2014, Helsinki, Finland.

Nye, S. (2011). "Headphone—Headset—Jetset." Dancecult: *Journal of Electronic Dance Music Culture.* 3.1. 64–96.

O'Donohue, John. (1999). *Eternal Echos: Celtic Reflections on Our Yearning to Belong.* New York: Harper Perennial.

Ofoghi, B., J. Zeleznikow, C. Macmahon, and M. Raab. "Data Mining in Elite Sports: A Review and a Framework." *Measurement in Physical Education and Exercise Science* 17.3 (2013): 171–186.

Ondaatje, Michael. (2002). *The Conversations: Walter Murch and the Art of Editing Film.* New York: Random House.

O'Reilly Media. (2012). Big Data Now: Current Perspectives from O'Reilly Media. Cambridge, MA: O'Reilly Publishers.

Pfadenhauer, M., (2009). "The Lord of the Loops. Observations at the Club Culture DJ-Desk." *Qualitative Social Research.* 10. 3, p. 1–17. http://www.qualitative-research.net/index.php/fqs/article/view/1338

Pfanner, E. "A Move to Curb Digitally Altered Photos in Ads." *New York Times.* 27 Sept. 2009. Web. 28 Mar. 2012. http://www.nytimes.com/2009/09/28/business/media/28brush.html?_r=0.

Pfiffner, P. (2010). "Photoshop Turns 20." *Digital Arts* 8 Feb. 2010: n. pag. Web. Picture Perfect: Life in the Age of the Photo Op.

Pierce, T. (2009). Social anxiety and technology: Face to face communication versus technological communication among teens. Computers in Human Behavior. 25, 1367–1372. Doi: 10.1016/j.chb.2009.06.003

"Photoshopping: Altering Images and Our Minds." *BEAUTY REDEFINED.* Beauty Redefined Foundation, 12 Mar. 2014. Web. 20 Mar. 2014. http://www.beautyredefined.net/photoshopping-altering-images-and-our-minds/.

Postman, N. (1992). "Technopoly: The Surrender of Culture to Technology." New York: Vintage.

———. (1979). *Teaching as a Conserving Activity.* New York: Dell Pub.

Prasad, A. (2005). "Making Images/Making Bodies: Visibilizing and Disciplining through Magnetic Resonance Imaging (MRI)." *Science, Technology & Human Values.* 2. 291–316.

Reedijk, S., Bolders , A., and B. Hommel. (2013). "The Impact of Binaural Beats on Creativity." *Front Human Neuroscience* 2013; 7: 786. Published online, 2013 Nov. 14. doi: 10.3389/fnhum.2013.00786

Riis, S. (2010)"A Sense of Postphenomenology." *Sats* 11.1: n. pag. Web.

Robinson, P. (2000). *Within the matrix: A hermeneutic phenomenological investigation of student experiences in web-based computer conferencing.* Retrieved 7/11/11 at http://otal.umd.edu/~paulette/Dissertation/finalmainmenu.htm

Rodowick. D. (2001). *Reading the Figural, or, Philosophy after the New Media.* Durham, NC: Duke University Press.

Rosenberger, Robert & Peter-Paul Verbeek. (2015). Postphenomenological Investigations: Essays on Human-Technology Relations (Eds). Lanham, MD: Lexington Books.

———. (2014). "A Postphenomenological Field Guide." *Postphenomenological Investigations: Essays on Human–Technology Relations.* Lanham, MD: Lexington Books

———. (2012). "Embodied Technology and the Dangers of Using A Phone While Driving." Phenomenology and Cognitive Science. New York: Springer Science & Business Media.

———. (2011). "Case Study In The Applied Philosophy of Imaging: The Synaptic Vesicle." *Debate, Technology & Human Values.* Thousand Oaks, CA: Sage Publications.

———. (2008). "Perceiving Other Planets: Bodily Experience, Interpretation, and the Mars Orbiter Camera." Human Studies. Vol 31(1). 63–75.

Rothenberg, D. (1993). *Hand's end: Technology and the Limits of Nature.* Berkeley: University of California Press.

Ruckenstein, M. (2014). "Visualized and Interacted Life: Personal Analytics and Engagements with Data Doubles." *Societies* 4.1 (2014): 68–84.

Scharff, B. and V. Dusek. (2013). *Philosophy of Technology: The Technology of Condition-An Anthology* (2nd ed.) Hoboken, NJ: Wiley & Blackwell.

Schrag, C. (1988). *Communicative Praxis and the Space of Subjectivity.* Bloomington: Indiana University Press.

Schull, N. (2016). *Keeping Track: Personal Informatics, Self-Regulation, and the Data-Driven Life.* New York: Farrar, Straus, and Giroux.

Schutz, A. (1973). *A Theory of Consciousness.* New York: Philosophical Library, 1973.

Seigworth, G. (2005). "From Affection to Soul." *Gilles Deleuze: Key Concepts.* New York: Acumen. 159–169.

Selinger, E. (2006). *Postphenomenology : A Critical Companion to Ihde.* Albany: SUNY Press.

Smoreda, Z. (2002). Communication technology and sociability: Between local ties and the global ghetto. The Machines that Become Us. Transaction Publishers. New York. 1–12.

Sobchack, V. (1992). *The Address of the Eye: A Phenomenology of Film Experience.* Princeton, NJ: Princeton University Press.

Sokolowski, R. (1999). *Introduction to Phenomenology.* Cambridge University Press.

Solberg, R. (2014). "Waiting for the Base to Drop." *Dancecult: Journal of Electronic Dance Music Culture.* 6.1. 61–82.

Spicer, R. (2014). "Long-Distance Caring Labor." *Techné: Research in Philosophy and Technology.* 18.1. 102–116.

Svenaeus, Fredrik. (2001). *The Hermeneutics of Medicine and the Phenomenology of Health: Steps towards a Philosophy of Medical Practice.* Dordrecht: Kluwer Academic.

Swan, J. (1990). *Sacred Places: How the Living Earth Seeks Our Friendship.* Santa Fe, NM: Bear & Company Publishing.

Tagg, P. (2015). *Music's Meanings: A Modern Musicology for Non Musos.* New York: The Mass Media Music Scholars' Press.

Taylor, M., and E. Saarinen. (1994). *Imagologies: Media Philosophy.* New York: Routledge.

"The Top 25." *Pew Research Center's Journalism Project RSS.* Pew Research Center, 08 May 2011. Web. 1 June 2014. http://www.journalism.org/2011/05/09/top-25/.

Thorson, E. (2008). "Changing Patterns of News Consumption and Participation: News Recommendation Engines." *Information, Communication & Society.* 11.4. 473–489.

Tiles, M., and H. Oberdiek. (1995). *Living in a Technological Culture: Human Tools and Human Values.* London: Routledge.

Turkle, S. (2011). *Alone together: Why we expect more from technology and less from each other.* New York: Basic Books
———. (1995). *Life on the Screen: Identity in the Age of the Internet.* New York: Simon & Schuster.
USA Triathlon. "Demographics." Accessed August 29, 2015.
Van Den Eede, Y. (2015). Tracing the Tracker: A Postphenomenological Inquiry into Self-Tracking Technologies. *Postphenomenological Investigations: Essays on Human-Technology Relations.* Lanham, MD: Lexington. 143–158.
Van Den Eede. Y. (2013). *Amor Technologiae: Marshall McLuhan as a Philosopher of Technology—Toward a Philosophy of Human-Media Relations.* Brussels: Brussels University Press.
van Manen, M. (1990). *Researching lived experience: Human science for an action sensitive pedagogy.* Albany: SUNY Press.
Van Veen, T. and Attias B. (2011). "Off the Record: Turntablism and Controllerism in the 21st Century (Part 1)." *Dancecult: Journal of Electronic Dance Music.* 3.1. DOI: 10.12801/1947-5403.2011.03.01.08.
Verbeek, P. (2015). Towards a Theory of Technological Mediation: A Program for Postphenomenological Research. *Technoscience and Postphenomenology: The Manhattan Papers.* Lanham, MD: Lexington Books. 2015. 189-204.
———. (2009). "Ambient Intelligence and Persuasive Technology: The Blurring Boundaries between Human and Technology." *NanoEthics* Vol. 3(3). 231–242.
———. (2006). "The Morality of Things: A Postphenomenological Inquiry." *Postphenomenology: A critical companion to Ihde.* Ed. Evan Selinger. Albany: SUNY Press. 117–128.
———. (2005a). *What Thing Do: Philosophical Reflections on Technology, Agency, and Design.* Translated by Bob Crease. University Park: The Pennsylvania State University Press.
———. (2005b). "Artifacts and Attachment: A Post-Script Philosophy of Mediation." *Inside the Politics of Technology: Agency and Normativity in the Co-Production of Technology and Society.* Amsterdam, Netherlands: Amsterdam University Press. 125–146.
Wajcman, J. (1991). *Feminism Confronts Technology.* University Park: Pennsylvania State University Press.
Wellner, G. (2015). Historical Variations and the Cellular Age. *Technoscience and Postphenomenology: The Manhattan Papers.* Lanham, MD: Lexington Books. 39–56.
———. (2015). *A Postphenomenology Inquiry of Cell Phones: Genealogies, Meanings, and Becomings.* Lanham, MD: Lexington.
———. (2015). "The Quasi-Face of the Cell phone: Rethinking Alterity and Screens." *Human Studies.* New York: Springer. September 2014, 37.3. 299–316.
Whyte, K. (2015). "What is Multistability: A Theory of the Keystone Concept of Postphenomenological Research." *Technoscience and Postphenomenology: The Manhattan Papers.* Lanham, MD: Lexington Books.
Winner, L. (1986). *The Whale and the Reactor: A Search for Limits in an Age of High Technology.* Chicago: University of Chicago Press.
Yamamiya, Y., T. Cash, S. Melnyk, H. Posavac, and S. Posavac. (2005). "Women's Exposure to Thin-and-beautiful Media Images: Body Image Effects of Media-ideal Internalization and Impact-reduction Interventions." *Body Image* 2.1: 74–80. 1954.
Zhok, A. (2012). "The Ontological Status of Essences in Husserl's Thought." *New Yearbook for Phenomenology and Phenomenological Philosophy.* Vol XI. 99–130.

Index

3D rendering, 3, 21
3D emersive technologies, 27
3D graphics, 53
4K ultra screen, 50

a priori, 61, 66
Aarseth E., 18, 173
Abram D., 173; on the gaze, 152; on perception, 10, 51; on presence, 65
actants, 90
actional-perceptual, 82
Adatto K., 115, 173
aerodynamics, 155
aesthetic, 18, 30, 60, 109, 115–117, 122, 123, 170; aesthetic distance, 56
affect, 42
alienation, 74
American Medical Association (AMA), 118
analytics, 127, 128, 131, 143, 159
Annenberg Media Exposure Research Group, 6
Anton C., xii, 67, 73, 78, 87
apparatus, 49, 65, 70, 72, 85, 125
apprehensions, 95, 150; apprehended, 95
architecture, 32
artifact, 5, 25, 31, 32, 41, 54, 56, 71, 104, 112, 115, 125, 127, 128, 132, 133, 152; material artifact, 5, 104, 128; non-neutral artifact, 56; technological artifact, 4, 30, 88, 106, 127

atmospheric, 87
Attias B., 102, 173, 180
Audio, 64, 70, 85, 99, 103; amplification, 85, 105; amplify, 105; amplitude, 17; audio as "three dimensional, interactive, synesthetic, 63; Audio broadcast streams, 129; Audio buzzers, 21; Audio content, 63; audio editing, 108; audio equipment, 94; audio processor, 101; audio professionals, 69, 70; audio technology, 92, 102; audio tracks, 68, 71; audio wave, 17; cart machine, 92; cassette player, 79; CD player, 80, 91; digital audio, 88; digital audio workstation, 106; flavored, 71; headphones, 50, 64, 66, 68, 79–81, 84, 85, 87, 98, 106, 150, 156; hearing loss, 97; hearing music, 85, 98; I-Doser/iDosing, 97; mp3, 68, 80, 84, 94, 106, 108, 129; my audio file, 93; noise canceling headphones, 86; out of tune, 15; personal audio devices, 91; plugged, x, 72, 80, 94, 96; plugged into a digital media device, 82; plugged into something, 85; plugged into the body, 81; plugged-out, 88, 89; plugged-in, 78, 85, 93, 97, 98; streaming audio apps, 89; transcribed audio, 36; unplugged, 85; videogame soundtracks, 102; visual and audio waves, 17
Augmented City, 51, 178

Index

augmented reality, 43, 50, 58, 60, 61, 119
aural, 64, 74, 90, 91, 95, 97, 115
authenticity, 67, 74, 87
Autodesk, 59
auto play, 68
aware, 34, 72, 93; awareness, 12, 13, 67, 73, 78, 82, 84, 91, 107, 151; unaware, 56, 78, 99, 128

balance, 64, 72, 73, 74, 78, 161, 162, 166, 170
Balsamo A., 125, 173
Barnhart R., 9, 36, 51, 54, 65, 69, 71, 173
beats, 97, 102, 104, 105, 108, 110, 111
Beck U., 38, 173
Being, ix, 4, 5, 7, 9, 10, 11, 13, 14, 16n1, 25, 36, 42, 44, 52, 57, 58, 59, 63, 64, 65, 67, 73, 78, 81, 84, 89, 95, 98, 121, 123, 160; being in, 74, 171; being in the other, 59; being-in-the-world, 11, 12, 13, 16, 27, 30, 41, 44, 66, 74, 84, 85, 90, 91, 93, 160, 171; being with, 64, 67, 92; being-in-place, 73; being-in-the-world with technology, 12, 26, 27, 37, 161; being-uploaded-into-the-world, 67; being-with-other-being-toward-world, 67, 73, 78, 87; being-with-others-being-toward-self, 78; being(s) in the midst of our technologies, 4, 7; embodied being, 57; field of being, 13, 84, 89, 98; monitor-world junction, 52; primordial sense-of-being-in-the-world, 13, 84; relatedness-to-Being, 14; sense of being-in-the-world, 13; sonorous being-with-others-being-toward-world, 67, 73, 87; Technological Other, 27, 57; tool being, 51; well-being, 109; world-as-expression, 43
Berendt J., 67, 68, 74, 173
between-ness, 25
bias, 23, 34; content biases, 31; ideological biases, 31; intellectual and emotional biases, 31; political biases, 31; sensory biases, 31; social biases, 31
bifractured, 19
binary, 17
binaural, 97
Blok V., 169, 170, 174
Bolter J., 6, 23, 174

Borgmann A., 6, 9, 174
Bostrom N., 121, 174
bracket, 7, 12
Bradbury R., 81, 174
Brey P., 53, 174
bricolage, 4, 7, 19, 150
broadband, 18
Broadcast Education Association, xii
Brugioni D., 115, 174
Buckingham D., 22, 174
Bull M., 66, 89, 91, 174

Cage J., 109, 174
case study, 19, 30, 34, 38, 44, 49, 79
Casey E., 61, 68, 73, 174
Cash T., 116, 174
chunking, 14
co-constitution, 10, 14, 15, 27, 30, 52, 57, 104, 128
co-shaping, 3, 10, 27, 38, 50, 103, 104, 133, 156, 160, 166, 167, 170, 171
code of ethics, 118
cognitive technologies, 3; cognitive functions, 90; cognitive processes, 97; cognitive psychology, 14; cognitive science, 143
collaboration, 101, 102, 103, 107, 110, 170
communal, 96
communication, 3, 6, 17, 20, 29, 30, 34, 39, 40, 50, 69, 73, 82, 83–84, 85, 87, 89, 93, 94–95, 95, 96, 97, 98, 170; communication technology, 50
computer, xi, 17, 18, 19, 23, 26, 31, 32, 44, 50, 51, 54, 58; Adobe Photoshop™, 115, 116; Adobe software, 116, 118; Apple brand, 21, 77; Apple iPhone, 21, 92; Apple iPod, 84, 92; Apple Watch, 21, 155; application developers, 30; application/app, 18, 60, 89, 136, 142; automation, 20, 92, 125; bike computer, 159, 162, 163, 165; compression rates, 50; computer engineering, 32; computer manipulation, 74; computer-generated graphics, 60; computerization, 12, 17; computerized bike fits, 155, 157, 158, 159; Excel, 135, 139; floppy disc, 53; human-computer relation, 60; human/computer connection, 59; inhuman, 43, 121, 122, 123, 125; interface, 11, 20,

53, 56, 57, 59, 60, 106, 108, 112, 133; Internet access, 142; Internet of things, 4, 6; Internet radio, 89, 91; machine, 4, 31, 44, 56, 106, 108; machine/computer, 53; mixed reality application, 50; music application, 89, 92, 112; networks, 7, 11, 18, 19, 24, 129, 131, 132, 148; nonhuman, 90; photo application, 115; tablets, 17, 20, 82; ubiquitous, xi, 3, 4, 20, 30; web application, 130; website, 68, 98, 102, 104, 107, 108, 110
connective, 80
construction of images of women, 116, 125
consumer, 5, 21, 22, 27
consumption, 3, 22, 80, 127, 132, 141
context, 9, 12, 16n1, 32, 39, 42; changing the context, 124; contemporary contexts, ix, xi; content and context, 118; context dependent, 133; contexted lifeworld, 149; cultural context, 39, 104, 134, 149; historical context, 42; non-neutral contexts, 40; situated within a surrounding context, 149, 151; socio-cultural context, 160; technological context, 58; use context, 18, 20; varied contexts, 160
Continental European tradition, 44
copyright infringement, 111–112
Couch L., 17, 174
Couldry N., 23, 174
Council of Better Business Bureaus in the U.S. National Advertising Division, 118
Crease R. P., 8, 45n1, 124, 174, 175
Creativecow.net, 69
Creeber G., 18, 174
critical theory, 16
critical-cultural studies, 15
Csikszentmihalyi M., 36, 174
Cubitt S., 18, 174
Culkin J., 5, 174
culture, ix, 5, 19, 38, 39, 40, 53, 54, 65, 66, 78, 80, 86, 90, 92, 93, 99, 101, 111, 112, 116, 125, 126, 128, 147, 158; audio culture, 92; British sound-system culture, 104; cultural anthropologists, ix, 106; cultural-hermeneutic, 149; EMC Culture, 103; iCulture, 92;

material culture, 38; message board culture, 108; on pluraculture, 150; pluralizing of cultures, 40; pop culture, 84; technology and culture, 88; youth culture, 72, 79. *See also* digital; DJ
curating, 142; as curation, 142
cyberspace, 53, 144; Barlovian cyberspace, 53
cyborg intentionality, 27

data, 6, 17, 18, 36, 60, 63, 68, 128, 129, 130, 131, 132, 133, 134, 135, 137, 139, 141, 166, 167; big data, ix, 18, 129, 131, 135, 136, 155; big data collection, 18; cloud, 117, 132, 135, 136; cloud storage, 135; data cleansing, 132, 136, 139; data manipulation, 131; data mining, ix, 127, 128, 130, 131, 132, 133, 134, 136, 138; data scraping, 139; data visualizations, 130, 138; data-driven journalism, 138; data-centric journalism, 130; data driven investigations, 130; dirty data, 139; keyword, 141, 146; metadata, 116; metatags, 148; Raw, 8, 105, 139, 158; raw data capture, 139; scraped, 133, 136, 139; site-tracking studies, 143; spam, 146; spider, 144, 146; tracking, 106; triangulating data, 129; webcrawler, 146
Davidson D., 53, 175
de Botton A., 143, 150, 175
death, 15, 64, 74, 123; modeling death, 123
Dekker A., 102, 175
Deuze M., 12, 175
device, xi, 3, 5, 6, 13, 17, 18, 20, 21, 22, 26, 43, 51, 54, 57, 63, 79, 80, 81, 85, 89, 94, 97–98, 98, 128, 147, 149, 152; aesthetic devices, 54; bring your own device (BYOD), 42; device in hand, 149; deviced sound, 85; digital media devices, 26; individuality of devices, 58; journalist + device + web + interface+ data, 133; medium/device, 25; one device experience, 89; personal audio device, 66; personal pocket device, 79; self-tracking device, 130, 155; ubiquitous devices, 38, 40, 42
Dewdney A., 19, 175

diagnostic, 155, 156, 157; as diagnostic bike fit, 159; as diagnostic technology, 37
Digfit, 157
digital, 4, 6, 17, 135, 136, 141, 142; algorithm, 111, 132, 135, 142, 144, 146, 148; digital code, 139; digital culture, 12; digital design, 16, 21; digital detox, 169; digital dirt, 130; digital dwelling, 68, 72, 73, 74; digital ecosystem, 137; digital images, 50, 115; digital listening, 65, 77, 89; digital media, xi, 3, 5, 6, 12, 16, 17–19, 19–22, 22–25, 26–27; digital presence, 65; digital radio station app, 82; digital sound, 37, 44, 63, 63–69, 64, 71, 74, 77–81, 88, 89–94, 101–104; digital space, 59; digital-alteration, 118; digitalization, 12; digitally enhanced, 65, 67; digitally mediated world, 4, 10; innovators, 32; non-digital sounds, 74
directionality, 116
disconnectedness, 66
discursive, 23, 125
display technology, 50
DJ, ix, 101, 103, 104, 106, 108, 110, 111; DJ culture, 102; DJ Merc, 107–111, 113n2; DJ performance, 80; DJ performance, 80, 102; DJ-software-live music venue, 81; DJ-software-sound website, 81; DJcity, 110, 111; Jamaican DJ cues, 105
Donath J., 44, 175
Dorfman E., 12, 175
Drake J., 159, 177
dubstep, 101–102, 103–105, 106; American East Coast Rave Scene, 103; beat mix, 108; brostep, 104; Collabinator, 111; drum machine/beatbox, 106; Dub Turbo, 105; dubstep nation, 101, 102, 112; final mix, 106; glitched sequences, 108; griminess, 107; hooks, 102, 106, 111; live mixing, 104; mini mix, 108; mix of digital components, 18; mix-tape, 79; mix(ed), 71, 101, 106, 110, 111; mixed reality applications, 50; mixes, 102, 103, 104, 105, 110; mixing, 70, 71, 80, 101; mixing on the fly, 104; newbstep, 108; oversharing, 111; post-dubstep, 104; purple dubstep, 104; remix, 101, 104, 105, 110; sampling, 102, 106, 111; SoundCloud, 92, 104, 108, 111; sub base, 105; sub-woofers, 105; subnav, 104; synthesizers, 102, 106; underground, 102, 130, 134; unmixed, 69; wobble bass/web, 105
duplication, 63, 69, 71
Dusek V., 8, 175
dwelling, 27, 61, 63, 64, 65, 68, 72, 72–74, 77, 90; dwelling place, 72, 73
Dyson F., 63, 175

earbuds, ix, 50, 64, 68, 77, 78–80, 81–84, 88; earbud etiquette, 81
economies of scale, 161, 165
eco-focus, 56
editing flatbed, 60
efficient, 18, 147; efficiency, 20, 155; efficiently, 20, 147, 148; sustainable power output and efficiency, 155
Eisenstein S., 8, 175
electrodes, 60
elite amateur athlete, 155, 156, 157, 164, 166, 167, 170; amateur, 126, 155, 156; amateur athletes, 156, 157, 158, 159; amateur multisport competition, 156; amateur triathletes, 160, 161, 164, 166, 167; final output, 155, 162, 167; Ironman, 156, 159, 161, 162, 165, 166, 167; marathon, 78, 159, 161, 162, 163; online comparison sites, 159; online mapping sites, 159; power, 167, 170; power numbers, 166, 167; power output, 155, 162, 167; power wattage meter, 159, 167; practice, 14, 17, 30, 32; sports medicine, 155; sports physiology, 155; *Strava* , 159; triathlete, 155–157, 158; triathlon, 156, 158, 159; USA Triathlon, 157, 161; VO2Max, 164, 167; wattage, 159, 167
embedded presence, 11
embodiment,: absent, 44, 81; absent body, 51, 59, 60; absorbed, 36, 56, 63, 77–78, 91; absorbed body, 51–52, 56, 59; alien/zombie embodiment, 123; as lived time, 10, 30; beauty, 117, 118, 119, 123, 126; Body One and Body Two, 42;

body schema, 13, 25; corporeal (lived body) schema, 13; corporeal, 10, 13, 25, 35, 36, 42, 59, 84, 85, 109, 167; corporeality, 25, 35, 36, 59; corporeally real and virtual, 10; disembodied, 13, 84; embodied, 4, 10, 13, 21, 26, 39; flesh, 83, 84; haptic feedback, 60; haptic pen, 58; haptic sensors, 21; heart rate monitor, 156, 158, 163, 165, 167; heart-rate, 86, 162, 163, 164, 167; heartbeat, 91, 155; lived body, 10, 36, 42, 56, 93, 99, 156; lived experience, 12, 33, 34, 35, 36, 44, 60, 89, 90; lived relation, 10, 36; lived space, 10, 36, 51; motility, 13, 81, 84; phenomenal body, 156; touch screens, 59, 60
empirical investigation, 30
empirical turn, 41
empirically oriented, 32
endistancing, 66, 90
enframing, 55; *Das Gestell*, 55
engineering psychologists, 21
enhanced, 119, 121, 156; enhanced performance, 156
entrepreneurs, 32
environments, 3, 105, 133, 157, 158; digital media environment, 4; environmental model, 134; laboratory environment, 164; mediated news environment, 149; smart environments, 4, 30, 39, 50; urban environments, 90
erfahrung, 13
ergonomics, 20, 155
erlebnis, 13
essence, 16n1, 32, 33, 34, 35, 105; essence and existence, 13, 84; essence of "the thing itself", 33; essence of technology, 9; essence of the question, 35; essences or variational elements, 107; structures or essences, 41; study of essences, 33; "whatness" or essence, 14
etiquette, 81, 91, 92, 93, 103
etymology, 33, 71, 130, 169
euphoric, 109
Evens A., 135, 175
existential, 34, 36, 57, 84
experience error, 79

face, xi, 10, 51, 57, 59, 82, 118, 121, 125; as face-to-face, 6, 57; as face-to-screen, 57, 149, 152; inter-face, 21; quasi-face, 57; quasi-faceness, 57
Fahrenheit 451, 81
feedback, 18, 60, 105, 110, 111, 128; as feedback loop, 68, 104, 106, 108
figure 9.1, 119
figure 9.2, 120
figure 9.3, 121
figure 9.4, 122
figure 9.5, 124
figure against the ground, 10, 149, 151
file sharing, 92
film, 6, 23, 52, 55, 58, 60, 70, 89, 105, 123, 128
film noir, 123
flow, 13, 23, 36, 92, 150
foley, 70
Foltyn J., 122, 123, 175
Ford S., 18, 19, 177
Forss A., 39, 40, 175
frame, 24, 31, 49, 54–55, 57, 60, 151; as frame—screen—window, 58, 59; as framework, 5, 9, 29, 30, 32, 33, 37, 38, 39, 44, 52
Friedberg A., 54, 55, 59, 60, 175
Friesen N., 23, 175
Friis J. K. B. O., 8, 45n1, 113n2, 175
fringe gadget, 85
fuel, 110, 158, 159, 170
functionalism, 30

Gadamer H., 7, 10, 12, 15, 33, 34, 35, 36, 56, 73, 175; on being, 73; on intellectual freedom, 7; on perception and being-in-the-world, 15, 73; on phenomenologists, 33–34; on spectators, 56; on the things themselves, 33
gait studies, 17
Gallagher S., 33, 175
gaming, 13, 17, 36, 41, 85; gaming console, 17, 20
Gant S., 129, 175
garden, 11, 86
gaze, 10, 11, 51, 107, 123–125, 152; fiction genres, 123; gaze into, 55; glassy-eyes gaze, 122; human gaze, 56; mediated

gaze, 125; medical gaze, 125; MR radiological gaze, 125; Splinter genres, 112
genre, 88, 99, 101, 102, 104, 105, 106, 108; dubstep genre, 101; music genre, 107; subgenre, 102, 104, 106
geographic, 40, 128, 129, 137; geographical, 129, 147; GPS, 60; GPS watch, 158, 162, 163, 164, 167
Gergen K., 67, 78, 175
gestalt, 40, 54, 132, 151; praxical gestalt, 106
Ghosh M., 13, 175
GIMP, 116
Gladwell M., 19, 175
GNU Image Manipulation Program, 116
Goeminne G., 8, 146, 175
Google, 141, 144, 145, 147, 148; Google culture, 146; Google friendly, 144; Google Glass, 53, 57, 60; Google News, 12, 144, 145, 146, 148, 149, 150; Google Scholar, 147; Google's user agent, 146; Googlebot, 144, 146, 148, 150
Gordon G., 159, 175
Green J., 18, 19, 177
Grusin R., 23, 174
Gunn J., 99, 175

Habermas J., 12, 175
Hall M., 99, 175
Hansen M.B.N., 18, 175
Haraway D., 7, 175
Harrison K., 115, 175
Hasse C., 175; on artifacts, 5; on multistability, 42, 127, 128
heads up displays (HUD), 21
Hefner V., 115, 175
Heidegger M., 8, 9, 10, 12, 14, 16n1, 33, 40, 45n1, 160, 176; On the essence of technology, 9
Herrera M., 105, 176
Hickman L.A., 30, 176
hit-making, 111
Holmstrom A. J., 116, 176
Honig E., 125, 126, 173
horizon, 13, 21, 25, 35, 43, 51, 55, 56, 59
Horowitz S., 63, 176
HTML, 139, 148

HUD (heads up display), 21
Hugg T., 23, 175
Hulu, 50
human-to-human connection, 15; human physiology, 155, 166; human science methodology, 33; human–technology connection, xi, 4, 6–10, 11–14, 15, 22–25, 27, 29–31, 32, 33, 34, 161, 163, 167, 169, 170, 171; human-media connection, 24; human–technology relations, 11, 30; human-technology-world, 4, 25, 30, 57; I-technology-world connections, 25; I-technology-world schema, 152; I-technology world model, 26; I-technology-world, 26; I-world perception, 25; instrument-human-world, 37; posthuman, 123, 124; super-human, 124; transhuman, 27, 121, 122, 124, 125
Husserl E., 10, 11, 12, 33, 40, 176
Hutchon K., 23, 178
hyperlink, 128, 150
hyper-sexualization, 116

iCat, 43
icon, 53; iconic, 53, 84, 92
Ihde D., x, xii, 176; on alterity, 26; on artifacts, 54; on embodiment, 167; on English longbow, 134; on hermeneutics, 82; on human-technology relations, 57, 90, 134; on humans and tools, 11, 15, 26; on "I-world" perception, 25; on synaesthetic perception, 99; on listening, 112; on macro- and micro-perception, 42, 82, 106; on multistability, 40, 82, 107, 134; on nonneutrality, 103; on presence, 9; Rosenberger on Ihde, 39; on scientific instruments, 11; on sound, 64, 78, 104; on technological texture, 4, 7; Verbeek on Ihde, 152
in-between, 21, 42, 52; in-between-ness, 26, 34
infographic, 128, 132
infrastructure, 18, 22, 137
internal customization, 20
interpreter, 33, 34
International Data Journalism Award, 130
interpretations, 5, 7, 32, 39, 40, 44, 97

iPad, 29
Irwin D. J. (cover photo), xiii
Irwin S., 3, 27, 57, 176; on Technological Other, 27, 57

Jacquard head dobby loom, xi
Jenkins H., 18, 19, 177
Johnson-Laird P., 14, 177
Johnson M., 53, 54, 59, 177
Jordan P., 20, 21, 177
Juslin P. N., 104, 177

Kaplan D. M., 8, 18, 177
Kaptelinin V., 177
Kelly Clarkson, 117
Kember S., 22, 177
Kovarkik B., 18, 177
Kurzweil R., 19, 177
Kuster J., 147, 177

Lakoff G., 53, 54, 59, 177
Langsdorf L., 37, 177
laptop(s), 50, 58, 60, 78, 89
Lawson C., 14, 177
LCD, 50
lebenswelt, 1.31
Leder D., 51, 177
Lee V. R., 159, 177
Lemmens P., 170, 174
lens-world junction, 52
Levin D. M., 13, 14, 63, 66, 67, 73, 74, 77, 78, 80, 81, 89, 90, 177
Liberati N., 8, 177
Liestol G., 18, 177
lifeworld, 3, 9, 12, 12–13
listening, 53, 64, 65, 66, 67, 68, 69, 73, 74, 78, 79, 80, 82, 85, 87, 97, 104, 108, 109, 129; 'seeing,' 'hearing,' 'sensing', 79; as soundscape, 82, 86, 88, 90, 92, 93, 94, 96, 98, 99; experience of listening, 109; functions of listening, 90; hearing, 10, 63, 66, 67, 68, 73, 74, 77, 79, 85, 90; human listening, 90; in-ear devices, 81; individual listening, 93; listening community, 108; listening self, 74; Listening to understand, 77; Mobile listening, 89; noise canceling, 84, 85, 86, 87; Portable listening, 89; Private listening, 80, 85; Sonorous,;

sonorous being-with-others-being-toward-world, 67, 73, 87; sonorousness, 67; Sony Discman, 99; Sony Walkman, 99; sound, 17, 63–67, 67; sound data, 63; sound waves, 17, 81; soundbed, 88; sweetening, 65; tuning in, 65; unlearned listening, 112; unmanipulated sound, 65
live mixing, 104
loadability, 67
Logic Studio, 105

magazines, 91, 115, 116, 123, 125, 148, 155, 157, 167
Magee H., 116, 177
mainstream media, 6
Manovich L., 6, 178
mass assemblage, 139
materiality, ix, 10, 22, 39, 56, 93; material artifacts, 5, 104; material "terrestrial" media, 6; material object, 56; material medium, 57; material technologies, 29; material technological artifact, 106; material world, 124
Matisse and Picasso, ix
Mazis G., 57, 98, 178
McLuhan E., 23, 178
McLuhan M., 23, 152, 178
media, ix, xi, 5, 6, 7, 10, 17, 22, 23, 24, 29; analog, 6, 17, 19, 23, 71, 79, 129; analog media predecessors, 19; analogue, ix; Arbitron, 129; as digitally mediated world, 4; as instrument-mediated perception, 52, 56; big media, 148; billboard, 17, 91, 115, 134; Blogger (website), 144; blogger, 3, 117; broadcast, xii, 17, 128, 129, 135, 148; cathode ray tube, 50; convergence, 16, 18, 169, 171; electronic media, 17; instrumental mediation, 25; interactive media, 22; intermediary, 25, 119; media content, 19; media gatekeeping, 150; media saturated world, 86; media studies, 6, 23, 24, 143; media-communication-informatics, ix; medialization, 22, 23; mediatic, 22, 23; mediation, 13, 16, 22, 23, 24, 25, 27, 29, 39, 52, 104, 106, 127, 132, 134, 146, 152, 166; mediatization, 16, 22, 23; mediator, 13, 17, 23, 30, 57, 84, 96,

119; new media, 6, 18, 22, 63, 144; remediation, 23; separated from device, 18; social media, 110, 115, 117, 119, 128, 129, 135; social sharing, 137; technological mediation, 27, 29, 39, 106, 132, 152, 167; See digital media
media ecology, 6, 15, 23
Media Ecology Association (MEA), xii
medical imaging, 37, 131, 132
Melnyk S., 116, 180
meme, 68, 117
Merleau-Ponty M., 10, 84, 95, 96, 178; being, 58, 95; being-in-the-world, 11, 81; body, 10, 59, 90, 96; corporeal schema, 84; experience error, 79; flesh, 10; habit, 95; horizon, 35; inhabiting, 59; perception, 98; sound, 87; "well in hand", 13
meta-key word, 148
metaphor,: as mine, 130; as frame, 54; as human-technology-world, 25; as orientational, 53; as machine, 44; machine/computer metaphor, 53; as technological texture, 4, 8; as weaving, 19, 36; as window, 55; the printing press, the computer and the television as metaphors, 31
mindfulness, 73
mine, 36, 130, 135, 136; "as mine", 92, 111, 130; a miner, 130. *See also* data mining
moral, 40, 170
Moran D., 33, 178
movie, 60, 102, 112, 153
Moviola, 60
MRI, 125
multidimensionalities, 38
multi platform, 50
multistability, 40, 69, 82, 106, 107, 116, 118, 126, 132, 133, 134, 144, 155
multiphrenic, 67, 78
multiplicity, 4, 41, 67, 78, 82, 98, 101, 107, 149
Munster A., 24, 178
Murch W., 74
music, 15, 26, 63, 64, 66, 68, 73, 78, 80, 84–85, 87, 89–91, 92, 93, 94, 98, 101–102; Auto-Tune, 101, 103, 104, 108; EMC, 102; jingle, 66; MIDI, 102, 103, 108; music consumption, 80; music listening, 77, 85, 89, 91, 97; music players, 20, 68, 84, 91; music-mixing tools, 106; music-wearing, 89; Vanilla Ice, 111
Music genres: Bass Big Room, 102; Chill Out, 102; Deep House, 102; drum& bass, 102; EDM, 101, 102, 103, 106, 107, 108, 112; Electro, 102; Future House, 102; Hard Style, 102; hip hop, 105; Techno, 105

Nagataki S., 156, 178
Nagel C., 20, 177
Napster, 110
Nardi B., 18, 177
National Communication Association (NCA), xii, 99n1
National Photographers Association, 118
NCR 551 printer memory buffer, xi
netizen, ix
new language, 107
news, 112, 116, 119, 127, 128, 131, 132, 134; aggregate, ix, 37, 44, 126, 134, 135, 137, 139, 145, 146; aggregate news, ix, 147, 148, 150, 151; aggregate news users, 150, 170; aggregated, 142, 144, 148, 149; Breaking news, 138; Buzzfeed, 147; Dig, 147; Journalism, 128, 129, 130, 131, 135, 137, 138; Mashable, 147; news sites, 137; news stories, 127, 135, 141, 144; newsgathering, 130, 132, 141, 142, 143, 152, 170, 171; newsworthy, 136; online newsreaders, 142; Reddit, 147; Slashdot, 142
non-neutral, 6, 30, 31, 40, 44, 56, 170, 171; non-neutrality, 5, 6, 12, 15, 20, 30, 63, 152
non-directional, 25
non-verbal, 93, 94
noob, 108
Nye S., 80, 178

O'Donohue J., 65, 178
O'Reilly Media, 129, 178
Oberdiek H., 160, 179
object of perception, 9; as object-correlate, 10

Ondaatje M., 74, 178
one device, 18, 41, 89, 90
ontological, 7, 14, 27, 34, 38, 52, 54, 64, 160; interrelated ontology, 9, 10, 14, 37, 41, 49; interrelated ontology, 10, 14, 15, 37, 49; interrelatedness, 9; ontological dimensionality, 84; ontology, 13, 14; pre-ontological, 12
Orson Welles, 66

Paredis E., 8, 175
participant observer, ix
PDF, 141
perception, 9, 10, 12, 13, 25, 32, 39, 40, 42, 43; as perceptualism, 40; body, 35, 36; cultural perception, 41; hermeneutic perception, 26; lived experience, 12; macro-perception, 38, 40, 42, 50, 53, 93, 107, 149, 170; micro-perception, 3, 25, 38, 40, 42, 43, 44, 49, 50, 89, 93, 98, 99, 107, 149, 150, 167; micro-perceptual, 3, 42, 77, 83, 89, 107, 167; microscopic, 116; microworld, 146; miroperceptually embodied, 83; object of perception, 8; of a garden, 11; perceptual mediation, 52; phenomenology, 11
performance, 101, 102, 105, 106, 107, 108, 109, 155, 159, 167, 170; athletic performance, 155; DJ performance, 80; Instrumental performance, 101; live performance, 101; maximum total performance, 155; multisport performance, 158; new performance experience, 106; peak performance, 167; perform, 103, 108; performance, 105, 158, 159, 167; performance artists, ix, 54; performance feel, 101; performed live, 108; performer, 107, 109; performing, 104, 110
personalized,– 66, 68, 77, 79, 81, 142, 145, 146, 148, 149, 150
Pew Internet and the American Life Project, 142, 148
Pew Research Center, 142
Pfadenhauer M., 102, 178
Philosophy of Communication, 15
Philosophy of Technology (PhilTech), 5, 9, 10, 24, 30

phones, 17, 116; cell, 15, 53, 57, 60, 78, 82, 89, 92, 131; electrophone, 79; mobile, ix, 128, 141, 142, 145, 150; smart, 4, 17, 20, 40, 58, 60, 142
phosphor-covered screens, 50
photo manipulation, 112, 115–118, 118–122, 126; photo cleansing, 116; photo-editing, 117; Photoshop, ix, 115; photoshopped, 116, 125; Photoshopping, 115, 116; touch up, 115, 116, 118; untouched photo, 119; zombie, 124. *See also* Adobe
playlist, 64, 77, 86, 87, 89, 105; automatic playlist generator, 112; personalized playlist, 80, 81. *See also* music
podcast, 64, 77, 78, 89
political, 22, 30, 31, 34, 38, 39, 40, 49, 52, 53, 133, 144, 150, 170
polymorphic, 42
portability, 67, 82
Posavac H., 116, 180
Posavac S., 116, 180
Postman N., 31, 178; benefits and deficits of technology, 5, 72; as burden and a blessing, 20; ideological bias, 31; metaphor, 31
Postphenomenological Research Group, xii
postphenomenology, ix, 5–10, 29, 31, 37, 44, 45n1, 102; applied phenomenology, 32; tool and artifact and instrument, 3, 22; artifacts, 127; case studies, 37; described, 38–39; empirical turn, 41; hermeneutic, 25, 26, 27, 34, 36, 42, 83; hermeneutic-phenomenology, 31, 34, 35, 41, 161, 167; interpretive framework, 29; linked with pragmatism, 30; multistability, 40, 69, 82, 106, 107, 116, 118, 126, 132, 133, 134, 144, 155; nonneutrality, 98; perceptual embodied experience, 78; phenomenology, 4, 9, 11, 14, 16n1, 29, 32, 33, 33–37, 57; postmodern, 45n1, 83, 167; pragmatist, 8, 30; reflexive practice, 7; semiotic(s), 15, 23, 51, 92, 97; sociological practice, 6; trajectories, 169; variations, 39
praxis: communicative praxis, 29; human praxis, 54; known praxis, 106, 150

Index

presence, 4, 15, 21, 65, 93, 98; embedded presence, 11; virtual presence, 22, 57; self presence, 59; digital presence, 65
production, 3, 6, 33, 50, 57, 58, 60, 80, 103, 105, 109; flows of, 23
public sphere, 82, 89

quasi-love, 26
quasi-hate, 26
questioning, 9, 33, 34

R&D (research and development), 38
radio, 6, 17, 66, 79, 81, 92, 110, 142, 153; digital radio station app, 82; early radio, 92; Internet radio, 89, 91; personal radio, 79; radio frequency identifications (RFID), 137; radio station, 77, 103; radio tower, 17; radiological gaze, 125; radiologist, radiological, 125; terrestrial radio, 92, 129; transistor radio, 79, 91
realism/irrealism, 133
relation, 10, 11, 16n1, 25, 26, 30, 36; as alterity, 26, 27, 57; as background, 98, 145, 151, 169; as embodiment, 10, 26; as hermeneutic, 25, 26; relationship, 4, 6, 9, 20, 24
remote, 52, 57
reoccurrent, 54
reproduce, 33, 69, 70, 71, 72
Ride P., 19, 175
Riis S., 37, 173
Robinson P., 160, 179
Rodowick D., 19, 179
Rosenberger R., 30, 179; on postphenomenology, 32, 37; embodiment, 40; on phenomenological analysis, 39; on multistability, 41
Rothenberg D., 43, 179
Royston M., 18, 174
Ruckenstein M., 22, 179

Saarinen E., 68, 179
saturated, 65, 73, 78, 86, 95
Scharff B., 8, 179
Schrag C., 29, 179
Schull N., 179; on self-tracking, 157
Schutz A., 12, 179
science-technology, ix

scratching, 106
screen, 49–52
search engine, 137, 141, 142; search engine optimization (SEO), 141, 148
Self magazine, 117
self-presencing, 59
self-select, 99, 145
self-tracking, 155–157, 160, 170; fitbit, 155, 157; gear biomechanics, 155; neural networks, 132; Neurotracking, 155; peak experience, 107; Quantified self, 130; Quantified Self Movement (QS), 155; Runkeeper, 157; sustainable power output and efficiency, 155; wearables, 50, 155
Seigworth G., 42, 45n1, 179; affect, 42; plane of immanence, 42
Selinger E., 54, 176, 179
sensory, 107, 149, 167, 170
shake, 57
sick beat, 110
simulacratic way, 128
Skrillex, 105, 107
Sloboda J., 109, 177
Smoreda Z., 6, 179
Sobchack V., 52, 53, 56, 92, 179
Society for Phenomenology and Human Sciences (SPHS), xii
Society for Phenomenology and Media (SPM), xii
Society for the Social Studies of Science (4S), xii
Society of Existential and Phenomenological Theory and Culture (EPTC), xii
socio-cultural, 4, 6, 44, 92, 98, 116, 160, 170
Sokolowski R., 33, 179
Solberg R., 109, 179
sonorousness 67
spectator, 124
Spicer R., 40, 56, 179
split screen, 55
spreadability, 19, 20, 23
stance, 150, 159, 160, 171
statistical, 111, 128, 129, 134, 136
structures, 11, 19, 31, 36, 39, 41, 69, 92, 107, 134
superheroes, 124

supermodel, 123
superficial, 21
supernatural, 109
surface,
Svenaeus F., 13, 72, 157, 179
Swan J., 179

Tagg P., 179
Taylor Swift, 118
Taylor M.,
teacher-scholar, xii, 7
techne, 8, 9, 27; craft, 19, 108, 127; art, craft and technology, 27
technological texture, ix, 4, 5, 6, 7, 12, 13, 15, 19, 21, 29, 33, 169–171
technology, 3–6, 8, 10, 13, 15, 16n1, 17, 20–21, 22, 24–27, 29–31, 32, 37–41, 43, 56, 57–58, 72, 77, 79, 82, 89, 92, 97, 101, 104, 107, 112, 115, 117, 125, 127–128, 131–132, 137, 141, 145, 146, 152, 155, 157, 160, 169–171; technoscience Studies, 37; technological directedness, 22; techno-utopian, 115; technologia, 25; technological entanglements, 24, 40, 42; technological instrumentality, 9; technological reciprocity, 49
television, 17
terrestrial, 6; terrestrial media, 6; terrestrial radio, 92, 129
The Odyssey, 64
things, 10, 14, 41; the things themselves, 33, 41
thinness, 35, 58
Thorson E., 147, 179
Tiles M., 160, 179
trajectory, 40, 91, 99, 103, 118, 133, 147, 150; trajectories, 8, 38, 40, 41, 83, 103, 107, 116, 133, 134, 169; of variations, 116
transcribed interview, 7, 15, 33, 34, 113n2, 161
transistor radio, 79, 91
translate, 11, 17, 51, 54, 57; translator, 57; translator tool, 54, 80
transparency, 8, 12, 52, 81, 84; transparency, partial, 56
Turkle S., 6, 60, 180
TV Displays, 50

twentieth century, 130
twenty-first century, 82

Ubersense, 159
UK, 104
Under Pressure, 111
Usability, 19, 20, 145; usability studies, 143, 150
user, 5, 7, 9, 10, 13, 20, 21, 22, 26, 115, 118, 119, 128, 133, 134, 141, 144; co-constitution, 14; tool user, 9, 10, 115; user and tool relationship, 24; user centered, 21; user friendly, 21, 22, 139, 146; user-and-tool relationship, 24; user's body, 13, 14

Van Den Eede Y., 180; on mediation, 24; on self-tracking, 157
van Manen M, 35, 180; on phenomenological writing, 33, 35; on phenomenology, 36; on the existentials, 36; on the reflexive practice of writing, 7
Van Veen T., 102, 180
variation, 7, 38, 39, 40, 42, 44
variational, 30, 38, 49, 107, 167
variational analyses, 39
variational intrepretations, 39
variational theory, 30, 39, 44, 49, 106
Verbeek P.P., 6, 25, 29, 119, 127–141, 152; on? mediation, 152; on artifacts, 128; on co-constitution, 133; on cyborg intentionality, 27; on directionality, 118; on macroperception, 149; on postphenomenology, 30, 37; on trajectory, 103
video games, 68
Vietnam War, xi
virtual, 53, 61, 157, 164; and canvas, 61; corporeally real and virtual, 10, 21; virtual and augmented, 43, 53; virtual opponents, 164; virtual presence, 22; virtual reality, 13, 27, 50, 52; virtual reality glasses, 52; virtual space/cyberspace, 144; virtual window, 53
vision, 35, 63, 64, 66, 68, 90
Volpe J., 119, 120, 121, 122, 124

War of the Worlds, 66

watch (to), 18, 50, 56, 58, 60, 97, 129, 142; watching, 50, 58, 59, 60, 129, 165
watch (wrist), 17, 20, 26, 57, 158, 162, 163, 165, 167. *See also* computer; GPS
Watergate Trial, 143
weave, 7, 10, 18, 19, 40, 43, 99, 169; weaver, 35; interweave, 29, 82; warp, ix, x, 19; weft, ix, 19, 171; weaver's loom, xi, 35
well in hand, 13, 84
Wellner G., 40, 57, 180
Whatness, 14
Whyte K., 40–41, 180
wifi, 18

Wikileaks, 130
window metaphor, 53, 54–56, 61, 170. *See also* metaphor
Winner L., 6, 180
workflows, 39, 92

Xfinity, 50

Yamamiya Y., 116, 180
YouTube, 50, 105, 111, 144

Zhok A., 34, 180
Zylinska J., 22, 177